航空类专业职业教育系列"十三五"规划教材

FEIJI JIEGOU XIULI ZUANYE YINGYU

飞机结构修理专业英语

主 编 魏 静 周宇静 吴成宝
副主编 朱 芳 刘传生

西北工业大学出版社

【内容简介】 本书共 16 课,涵盖了飞机结构专业维修的工具、设备、飞机结构和设施以及典型维修工艺、维修手册等方面的内容。每课由课文、词汇表、注释、练习和拓展阅读材料组成,课后的词汇表列出课文中出现的常用专业词汇,给出了专业的解释,并在全书的最后按照字母顺序归纳编排,便于读者查找和记忆。练习和拓展阅读材料一方面巩固所学的课文知识,另一方面也进一步拓展了读者关于该课文延伸的专业知识,确保读者可以掌握和运用所学的知识。

本书可作为高等院校飞机相关专业的专业英语课程教材,尤其适合于飞机结构修理、机电修理、客舱修理等专业的学生,而且也可以供飞机维修等专业的从业人员作为培训教材,也可以作为飞机维修的专业技术人员日常参考资料。

图书在版编目(CIP)数据

飞机结构修理专业英语/魏静,周宇静,吴成宝主编 . —西安:西北工业大学出版社,2016.11
航空类专业职业教育系列"十三五"规划教材
ISBN 978 - 7 - 5612 - 5136 - 2

Ⅰ.①飞… Ⅱ.①魏… ②周… ③吴… Ⅲ.①飞机构件—维修—职业教育—教材 Ⅳ.①V267

中国版本图书馆 CIP 数据核字(2016)第 271056 号

策划编辑:华一瑾
责任编辑:卢颖慧

出版发行:西北工业大学出版社
通信地址:西安市友谊西路 127 号　　邮编:710072
电　　话:(029)88493844　88491757
网　　址:www.nwpup.com
印　刷　者:兴平市博闻印务有限公司
开　　本:787 mm×1 092 mm　　1/16
印　　张:20.125
字　　数:491 千字
版　　次:2016 年 11 月第 1 版　2016 年 11 月第 1 次印刷
定　　价:48.00 元

前　言

　　随着我国经济的发展,中国民航运输业、航空制造业等也在这个阶段飞速发展,由此带来了中国的飞机维修行业高速发展。目前中国从事飞机大修业务的 MRO 数量也越来越多,各个航空公司的维修能力和维修深度也越来越深,飞机结构修理专业已经成为了飞机维修领域中极其重要的专业,而且从业人员的数量增加很快。由于目前绝大部分民航飞机都是进口飞机,维护人员必须要有相关的专业英语能力,为了满足飞机结构修理技术人员和管理人员的需求,我们编写了《飞机结构修理专业英语》一书。

　　本书的内容主要从飞机结构维修专业的角度出发,从结构修理的工具、紧固件、结构检测、飞机结构和设施、典型维修工艺和方法、结构修理手册等方面介绍飞机结构专业所涉及的专业知识,很多内容就来自于实际工作,并参考飞机维修手册,确保全书内容不仅仅满足于对飞机结构专业英语的学习,而且也包含一定的飞机结构修理的专业技术知识。

　　每课由课文、词汇表、注释、练习和拓展阅读材料组成,课后的词汇表列出课文中出现的常用专业词汇,给出了专业的解释,并在全书的最后按照字母顺序归纳编排,便于读者查找和记忆。练习和拓展阅读材料一方面巩固所学的课文知识,另一方面也进一步拓展了读者关于该课文延伸的专业知识,确保读者可以掌握和运用所学的知识。

　　本书第 1,3,4,15 课由魏静编写;第 10～13,16 课由周宇静编写;第 5,8,14 课由朱芳编写;第 6,7 课由刘传生编写;第 2,9 课以及词汇表由吴成宝编写;吴成宝负责参考资料汇总,魏静负责全书统稿和整体框架设计。

　　本书课文和拓展阅读材料精选自国外原版书刊或波音公司和空客公司提供的技术资料和飞机维修手册等内容,突出了飞机结构主要涉及的内容,专业性和实用性强,难度适宜,系统全面,既可作为高等院校航空相关专业的专业英语课程教材,也可作为飞机结构维修专业的从业人员学习的参考。

　　由于水平有限,书中难免存在疏漏,恳请各位专家和读者指正。

编　者

2016 年 8 月

目　　录

目 录

Lesson 1   Basic Knowledge

INTENSIVE READING

Arithmetic, Geometry and Trigonometry

Number Systems

Arithmetic uses real, non-negative numbers, which are also known as counting numbers, and consists of only four operations, addition, subtraction, multiplication, and division. In daily life, most people typically use a "base ten" or decimal system. However, another numbering system that is used in computer calculations is the binary, or "base two" system. The decimal system is based on ten whole numbers, often called integers, from zero to nine. The base two, or the binary system, only utilizes the digits zero and one.

Whole Numbers

Addition

The process of finding the total of two or more numbers is called addition. This operation is indicated by the plus (+) symbol. When numbers are combined by addition, the resulting total is called the sum.

Subtraction

The process of finding the difference between two numbers is known as subtraction and is indicated by the minus (−) sign. The number which is subtracted is known as the subtrahend, and the number from which the quantity is taken is known as the minuend. Subtraction is the reverse of addition

Multiplication

When a given number is added to itself a specified number of times, the process is called multiplication. The number multiplied is called the multiplicand, and the multiplier represents the number of times the multiplicand is added to itself. The multiplicand and the multiplier are called factors and the answer represents the product. Multiplication is typically indicated by an (×), (·), or in certain equations, by the lack of any other operation sign.

Division

Division is the reverse of multiplication. Division is a means of finding out how many times a number is contained in another number. The number divided is called the dividend, the number you are dividing by is the divisor, and the result is the quotient. Division is indicated by the use of the division sign (÷) with the dividend to the left and the divisor to the right of the sign. Division also is indicated in fractional form.

Signed Numbers

If zero is used as a starting point, all numbers larger than zero have a positive value, and those smaller than zero have a negative value.

Common Fractions

A common fraction represents a portion or part of a quantity. For example, if a number is divided into three equal parts, each part is one-third (1/3) of the number. A fraction consists of two numbers, one above and one below a line, or fraction bar. The fraction bar indicates division of the top number, or numerator, by the bottom number, or denominator.

Decimals

In a decimal, each digit represents a multiple of ten. The first digit represents tenths, the second hundredths, the third thousandths. Example:

0.5   is read as five tenths;
0.05   is read as five hundredths;
0.005   is read as five thousandths.

Percentage

Percentages are special fractions whose denominator is 100. The decimal fraction 0.33 is equivalent to 33 percent or 33%. Another way percentages are used is to determine a number when only a portion of the number is known.

Ratio and Proportion

A ratio provides a means of comparing one number to another. The use of ratios is common in aviation. One ratio you must be familiar with is compression ratio. Another typical ratio is that of different gear sizes.

A proportion is a statement of equality between two or more ratios and represents a convenient way to solve problems involving ratios.

Powers and Roots

When a number is multiplied by itself, it is said to be raised to a given power. The number of times a base number is multiplied by itself is expressed as an exponent and is written to the right and slightly above the base number. A positive exponent indicates how many times a number is multiplied by itself. For example, 3^2 is read "3 squared" or "3 to the second power". 2^3 is read "2 cubed" or "2 to the third power". A negative exponent implies division or fraction of a number. It indicates the inverse, or reciprocal of the number with its exponent made positive. For example, 2^{-3} is read "2 to the negative third power".

The root of a number is that value which, when multiplied by itself a certain number of times, produces that number. For example, 4 is the square root of 16; the cube root of 27 is 3.

Scientific Notation

Many engineering and scientific calculations involve very large or very small numbers. To ease manipulation and decrease the possibility for error, scientific notation is used. Scientific notation is based on multiplying a number by a power of ten. (Figure 1-1)

Positive Powers of Ten	Negative Powers of Ten
$10^0 = 1$	$10^{-1} = 0.1$
$10^1 = 10$	$10^{-2} = 0.01$
$10^2 = 100$	$10^{-3} = 0.001$
$10^3 = 1,000$	$10^{-4} = 0.000,1$
$10^4 = 10,000$	$10^{-5} = 0.000,01$
$10^5 = 100,000$	$10^{-6} = 0.000,001$
$10^6 = 1,000,000$	

Figure 1-1　Scientific notation

Geometry is the measurement of dimensions, areas, and volumes of geometric shapes, and is quite useful in aviation maintenance. In fact, it is geometry that allows you to calculate the displacement of a cylinder, determine the volume of a fuel tank, and calculate the surface area of a wing. On the other hand, trigonometry allows you to determine unknown lengths and angles of a triangle. In addition to aiding you when fabricating sheet metal, trigonometry plays a large part in the theory of alternating current.

Computing Area

The area of a surface is two dimensional and is expressed in square units. An area that is square and measures one inch on each side is called a square inch. This same relationship holds true for other units of measure such as square feet, square yards, square miles, and square meters.

The Rectangle

As you know, a rectangle is a four-sided plane. It is distinguished by having opposite sides of equal length, and four angles each equal to 90 degrees. The area (A) of a rectangle is found by multiplying its length (L) by its width (W), or $A = L \times W$ (Figure 1 - 2).

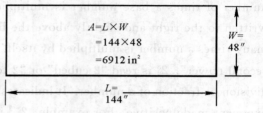

$$A = L \times W$$
$$= 144 \times 48$$
$$= 6912 \text{ in}^2$$

The Square

A square is a symmetrical plane in which all four sides are of equal length. The formula is sometimes expressed as the square of the sides or: $A = S^2$.

Figure 1 - 2　Rectangle's area

The Triangle

The triangle is a three-sided figure consisting of three angles whose combined measurement equals 180 degrees. Three basic types of triangles you should be familiar with are: the scalene triangle, the equilateral triangle, and the isosceles triangle. Triangles are further classified by the measurement of one angle. They are: the right triangle, the obtuse triangle, and the acute triangle. The formula for calculating the area of a triangle is one-half the base times the height (Figure 1 - 3).

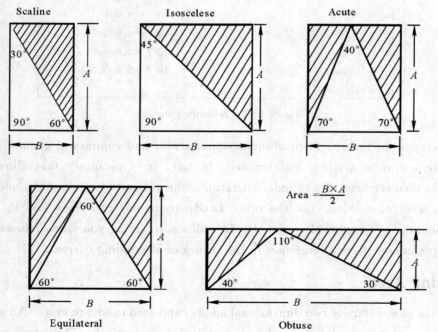

Figure 1 - 3　The shaded portion of each rectangle is equal in area to the triangle that is not shaded. The area of a triangle is calculated with the formula Area = $1/2 B \times A$

The Parallelogram

The parallelogram, like the rectangle, has opposite sides that are parallel and equal in length. However, the corner angles of a parallelogram are some measurement other than 90 degrees. The area of a parallelogram is calculated by multiplying the length by the height (Area=$B\times A$) (Figure 1-4).

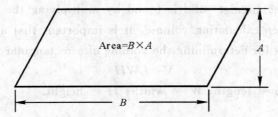

Figure 1-4   The parallelogram's area

The Trapezoid

A trapezoid is a four-sided figure that has one set of parallel sides. The trapezoid's area is equal to one-half the product of the base times the height. This is expressed with the formula: Area=$1/2(B_1+B_2)A$ (Figure 1-5).

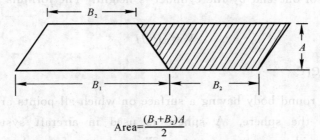

$$\text{Area}=\frac{(B_1+B_2)A}{2}$$

Figure 1-5   The trapezoid's area

The Circle

A circle is a closed figure bounded by a single curved line. Every point on the line forming a circle is an equal distance from the center. The circumference of a circle is found by multiplying the Greek letter pi (π) times the diameter, and the area is calculated by multiplying pi times the square of the radius.

$$\text{Circumference}=\pi D$$
$$\text{Area}=\pi R^2$$

Computing Volume

Solids are objects with three dimensions: length, width, and height. Having the ability to calculate volume enables you to determine the capacity of a fuel tank or reservoir, figure the capacity of a cargo area, or calculate the displacement of a cylinder. Volumes are calculated in cubic units, such as cubic inches, cubic feet, and cubic centimeters. However, volumes are easily converted to useful terms such as gallons.

Volume of a Rectangle

The volume of a rectangular solid is found by multiplying the dimensions of length, width, and height. When calculating volume, it is important that all measurements be in like terms. The formula for determining the volume of a rectangular solid is

$$V = LWH$$

Where: V = volume; L = length; W = width; H = height.

Volume of a Cube

A cube is a solid with equal sides. Its volume is calculated by multiplying one dimension by itself three times.

Volume of a Cylinder

A cylinder is a solid with circular ends and parallel sides. Its volume is found by multiplying the area of one end by the cylinder's height. The formula is expressed as see Figure 1 – 6

$$V = \pi R^2 H$$

Volume of a Sphere

A sphere is any round body having a surface on which all points are an equal distance from the center of the sphere. A sphere is used in aircraft systems for hydraulic accumulators and liquid oxygen converters. The volume of a sphere is determined by multiplying the cube of the diameter by a factor which is 1/6 pi, or 0.523,6. The formula is expressed as see Figure 1 – 7.

$$V = \pi D^3 / 6$$

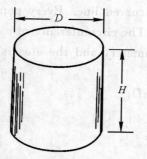

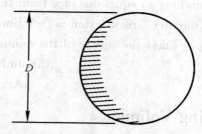

Figure 1 – 6　The cylinder's volume　　　　Figure 1 – 7　The sphere's volume

Trigonometry Functions

Trigonometry basically deals with the relationships that exist within a right triangle and is commonly used in the shop for sheet metal layout. Because trigonometry is a based on the ratio of the sides of a right triangle to one another, you must be familiar with how these ratios are derived. Figure 1 – 8 illustrates a right triangle with the sides and angles labeled for identification. Angle C is the right angle (90°).

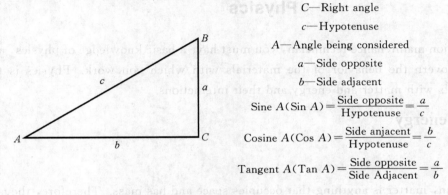

C—Right angle

c—Hypotenuse

A—Angle being considered

a—Side opposite

b—Side adjacent

$$\text{Sine } A(\text{Sin } A) = \frac{\text{Side opposite}}{\text{Hypotenuse}} = \frac{a}{c}$$

$$\text{Cosine } A(\text{Cos } A) = \frac{\text{Side anjacent}}{\text{Hypotenuse}} = \frac{b}{c}$$

$$\text{Tangent } A(\text{Tan } A) = \frac{\text{Side opposite}}{\text{Side Adjacent}} = \frac{a}{b}$$

Figure 1 – 8　Trigonometric relations of a right triangle

The Metric System

In the United States, the customary units of measurement include the English units of inches, feet, ounces, and pounds. However, the metric system is the dominant language of measurement in use today.

The metric system is built on decimal units. Each basic unit is divided or multiplied by ten as many times as necessary to get a convenient size. Each of the multiples has a definite prefix, symbol, and name. As a technician, you must be familiar with each of them (Figure 1 – 9).

Number	Prefix	Symbol	Scientific notation
1,000,000,000,000	tera	t	1×10^{12}
1,000,000,000	glga	G	1×10^{9}
1,000,000	mega	M	1×10^{6}
1,000	kilo	k	1×10^{3}
100	hecto	h	1×10^{2}
10	deka	dk	1×10^{1}
0.1	decl	d	1×10^{-1}
0.01	centi	c	1×10^{-2}
0.001	milli	m	1×10^{-3}
0.000,001	micro	μ	1×10^{-6}
0.000,000,001	nano	n	1×10^{-9}
0.000,000,000,001	pico	p	1×10^{-12}

Figure 1 – 9　Listed are the common prefixes, symbols, and multiples for basic metric quantities

There are six base units in the metric system. The unit of length is called the meter, and is approximately 39 inches. The metric unit of mass, or weight, is the gram. The unit of time is the second. The unit of electrical current is the ampere. The unit of temperature is the degree Celsius, formerly called degree Centigrade. The unit of luminous intensity is the candela. All other units of measurement in the International System of Units are derived from these six.

Physics

As an aviation maintenance technician, you must have a basic knowledge of physics, and the laws that govern the behavior of the materials with which you work. Physics is the science that deals with matter and energy and their interactions.

Matter and energy

Matter

By definition, matter is anything that occupies space and has mass. Therefore, the air, water, and food you need to live, as well as the aircraft you maintain, are all forms of matter. Matter may exist in one of three physical states, solid, liquid, and gaseous. For example, ice, water, and steam are all H_2O.

Energy

Energy is the capacity of an object to perform work. It is classified into two rather broad types, potential and kinetic. Potential energy is energy stored in a material and is divided into three groups: (1) that due to position, (2) that due to distortion of an elastic body, and (3) that which produces work through chemical action (Figure 1 - 10). When potential energy is released and causes motion, it is changed to kinetic energy. Kinetic energy is known as "energy of motion".

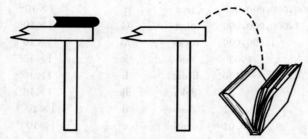

Figure 1 - 10　When the book is at rest it possesses potential energy because of its position; but as it falls, it has kinetic energy because of its motion

Units of Energy

In order to better understand energy, you must recognize the units with which it is expressed. The most common unit of measure of mechanical energy is the horsepower and is equivalent to 33,000 foot-pounds of work done in one minute. In the metric system the measure of mechanical energy is the Joule. For electrical energy the typical unit of measure is the watt. These units are used extensively in the study of machines and electricity.

Work, Power, Force, and Motion

Work, power, force, and motion are important concepts of physics. As an aircraft maintenance technician, you must understand these concepts and be able to use the associated formulas to fully comprehend simple machines like the lever, pulley, or gear.

Work

If a force is applied to an object and the object moves, work is done. The amount of work done is directly proportional to the force applied and the distance the object moves. In mathematical terms, work is defined as the product of force times distance. This is expressed by the equation: Work = Force $(F) \times$ Distance (D). If a force is applied to an object and the object does not move, no work is done.

In the English system, work is typically measured in foot-pounds. One foot-pound is equal to one pound of force applied to an object through the distance of one foot. In the metric system, the unit of work is the joule. One joule is the work done by a force of one newton acting through a distance of one meter.

Power

When determining the amount of work done, the time required to do the work is not considered. Power, on the other hand, does take time into consideration. Power is calculated with the formula

$$\text{Power} = (\text{Force} \times \text{Distance}) / \text{Time}$$

Power is defined as the time-rate of doing work. In the English system, power is expressed in foot-pounds per second, whereas the unit of power in the metric system is joules per second. Another unit of measure for power is the horsepower. Horsepower was first used by James Watt to compare the performance of his steam engine with a typical English dray horse. One horsepower is the amount of power required to do 33,000 foot-pounds of work in one minute or 550 foot-pounds of work in one second. The electrical unit of measure for power is the watt.

Mechanical Advantage

You now know that work is the product of a force applied to an object, times the distance the object moves. However, many practical machines use a mechanical advantage to change the amount of force required to move an object. Some of the simplest mechanical advantage devices used are: the lever, the inclined plane, the pulley, and gears (Figure 1 – 11).

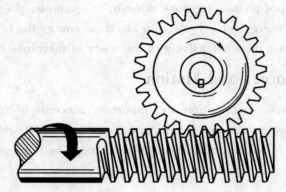

Figure 1 – 11　A worm gear system provides a very high mechanical advantage

Stress

When an external force acts on a body, it is opposed by an internal force called stress. The English measure for stress is pounds per square foot, or pounds per square inch. Stress is shown as the ratio

$$Stress = External\ force/Area\ of\ applied\ force$$

There are five different types of stress in mechanical bodies. They are tension, compression, torsion, bending, and shear.

Strain

As stated earlier, stress is a force within an object that opposes an applied external force. Strain is the deformation of an object that is caused by stress. Hooke's law states that if strain does not exceed the elastic limit of a body, it is directly proportional to the applied stress. This fact allows beams and springs to be used as measuring devices. For example, as force is applied to a torque wrench, its deformation, or bending, is directly proportional to the strain it is subjected to. Therefore, the amount of torque deflection can be measured and used as an indication of the amount of stress applied to a bolt.

Motion

Speed and Velocity

Speed and velocity are often used interchangeably. However, they are actually quite different. Speed is simply a rate of motion, or the distance an object moves in a given time. It is usually expressed in terms like miles per hour, feet per second, kilometers per hour, or knots. Speed does not take into consideration any direction. Velocity，on the other hand，is the rate of motion in a given direction，and is expressed in terms like five hundred feet per minute downward，or 300 knots eastward. An increase in the rate of motion is called acceleration and a decrease is called deceleration. Both acceleration and deceleration are measured in terms such as feet per second per second，or meters per second per second.

Vectors

A vector quantity is a mathematical expression having both magnitude and direction. Velocity is a vector quantity because it has both of these characteristics. Since all vector quantities have magnitude and direction，they can be added to each other. One of the simplest ways to add vectors is to draw them to scale.

Circular Motion

When an object moves in a uniformly curved path at a uniform rate，its velocity changes because of its constant change in direction. The force that pulls the spinning object away from the center of its rotation is called centrifugal force. The equal and opposite force required to hold the weight in a circular path is called centripetal force. Centripetal force is directly proportional to the mass of the object in motion and inversely proportional to the size of the circle in which the object travels.

New Words/Phrases/Expression

1. arithmetic[əˈriθmətik]　*n.*算术
2. geometry[dʒiˈɔmitri]　*n.*几何学
3. trigonometry[ˌtrɪɡəˈnɑmətri]　*n.*三角学，三角法
4. decimal[ˈdesɪml]　*n.* & *adj.*小数；十进制的，小数的
5. binary[ˈbaɪnəri]　*n.* & *adj.*二进制；二元的
6. multiplication[ˌmʌltɪplɪˈkeɪʃən]　*n.*乘法
7. division[dɪˈvɪʒən]　*n.*除法
8. fraction[ˈfrækʃən]　*n.*分数
9. proportion[prəˈpɔrʃən]　*n.* & *v.*比例；使成比例
10. power[ˈpauə(r)]　*n.*［数］幂，［机］功率
11. rectangle[ˈrektæŋgl]　*n.*长方形，矩形
12. square[skweə(r)]　*n.*正方形
13. equilateral triangle　*n.*等边三角形
14. isosceles triangle　*n.*等腰三角形
15. parallelogram[ˌpærəˈleləgræm]　*n.*平行四边形
16. trapezoid[ˈtræpəzɔɪd]　*n.*梯形
17. circumference[səˈkʌmfərəns]　*n.*圆周，周长

18. cube[kju:b]　n.立方体,立方
19. cylinder['sɪlɪndə(r)]　n.圆柱体,气缸
20. sphere[sfɪə(r)]　n.球体
21. volume['vɒlju:m]　n.体积;音量
22. kinetic[kaɪ'netɪk]　adj.运动的
23. work[wɜ:k]　n.功
24. lever['li:və(r)]　n.杠杆
25. pulley['pʊli]　n.滑轮
26. gear[gɪə(r)]　n.齿轮
27. stress[stres]　n.应力
28. strain[streɪn]　n.应变
29. velocity[və'lɒsəti]　n.速度
30. vector['vektə(r)]　n.矢量,向量
31. centrifugal[ˌsentrɪ'fju:gl]　adj.离心的
32. centripetal[sen'trɪpɪtl]　adj.向心的

Notes

(1)Trigonometry basically deals with the relationships that exist within a right triangle and is commonly used in the shop for sheet metal layout.

翻译:三角函数主要解决直角三角形内的边角关系,常常用于车间里金属板材的布局放样。

(2)For example, as force is applied to a torque wrench, its deformation, or bending, is directly proportional to the strain it is subjected to. Therefore, the amount of torque deflection can be measured and used as an indication of the amount of stress applied to a bolt.

分析:torque wrench 是力矩扳手;be directly proportional to 与……直接成正比(be inversely proportional to 与……成反比);it is subjected to 作为后置定语修饰 strain。

翻译:例如,在扭矩扳手上施加一个力,其变形或弯曲是与它受到的应变直接成正比的。因此,扭矩偏转量是可以测量的,并可以用来指示施加到螺栓上的应力的量。

Exercises

I. **Answer the following questions.**

(1)Which numbering system is used in computer calculations?

(2)How many operations does arithmetic consist of? What are they?

(3)List two uses of ratios common in aviation.

(4)What are geometry and trigonometry used for?

(5)Express the formulas to find the area of these shapes: a rectangle, a square, a triangle, a parallelogram, a trapezoid and a circle.

(6)Express the formulas to find the volume of these solids: a rectangular solid, a cube, a cylinder and a sphere.

(7)What English units do the customary units of measurement include in the United States?

(8)How many base units are there in the metric system? What are they?

(9)Matter may exist in one of three physical states. What are they?

(10)What three groups is potential energy divided into?

(11)What units are used for the measure of mechanical and electrical energy?

(12)What is the difference between the work and the power?

(13)How many different types of stress are there in mechanical bodies? What are they?

(14)What is the difference between speed and velocity?

Ⅱ. Write the following numbers as words.

0.43   0.532   0.002,8   0.280,0

Ⅲ. Put the following sentences into Chinese.

(1)A circle is a closed figure bounded by a single curved line. Every point on the line forming a circle is an equal distance from the center. The circumference of a circle is found by multiplying the Greek letter pi(π) times the diameter, and the area is calculated by multiplying pi times the square of the radius.

(2)Because trigonometry is a based on the ratio of the sides of a right triangle to one another, you must be familiar with how these ratios are derived.

(3)Energy is the capacity of an object to perform work. It is classified into two rather broad types, potential and kinetic.

(4)Many practical machines use a mechanical advantage to change the amount of force required to move an object. Some of the simplest mechanical advantage devices used are: the lever, the inclined plane, the pulley, and gears.

(5)Hooke's law states that if strain does not exceed the elastic limit of a body, it is directly proportional to the applied stress. This fact allows beams and springs to be used as measuring devices.

(6)When an object moves in a uniformly curved path at a uniform rate, its velocity changes because of its constant change in direction. The force that pulls the spinning object away from the center of its rotation is called centrifugal force. The equal and opposite force required to hold the weight in a circular path is called centripetal force.

(7)Centripetal force is directly proportional to the mass of the object in motion and inversely proportional to the size of the circle in which the object travels.

Ⅳ. Reading material.

Flying

Men have always wanted to fly like birds. Birds can fly easily because they are light; but men's bodies are heavier.

Men first went up into the air in balloons. These are big bags, and they are filled with gas. Hydrogen is a useful gas for balloons. It is lighter than air. Helium is also lighter than air, but it costs a lot of money. So balloons were (and are) usually filled with hydrogen.

Balloons have to fly with the wind as they have no engines to drive them against it. Later, men made airships. These were balloons with engines, but they were not round. They were long, and the engines were at the back. They were also filled with hydrogen and some of them caught fire because the hydrogen escaped and the engines heated it. Then the airship was completely burned in a few seconds.

Aircraft with wings now take people across the world. Powerful engines drive these machines across the sky. Some of the engines are like the engines of cars, but they are more powerful.

There is another kind of engine which we call the jet engine. An English engineer invented the jet engine. In May 1941 his new engine was fixed in an aircraft, and the aircraft flew quite well. At the same time the Germans were also building a jet engine; but neither country told the other, of course.

Jet engines are very powerful. Usually two, three or four are enough for an aeroplane; but some big aircraft have six. Anyone in a moving jet plane can feel the power of the engines. Jet planes can travel faster than sound. (Sound travels at about 1,100 feet a second. That is about 760 miles an hour.) As a flying jet plane leaves its noise behind it, we do not hear it until it has gone.

(1) According to the passage, balloons are usually filled with hydrogen because _____.

    A. it makes them go faster   B. it is cheaper

    C. it is heavier   D. it is lighter than helium

(2) Which of the following is not true?

    A. The first balloons that helped people travel in the air didn't have engines.

    B. The first balloons that people traveled with were round.

    C. Weather exerted a greater impact on balloons than it does on today's aeroplanes.

    D. Balloons could go to all the places as people wished, so they were very useful.

(3) Airships are different from balloons in that they _____.

    A. have engines   B. are longer

    C. don't have to be filled with gas   D. travel at a higher speed

(4) It is implied in the passage that it is most dangerous _____.

    A. to travel by balloon   B. to travel by airship

    C. to drive a car   D. to travel for a long distance

(5) With which would the author most probably agree?

    A. Modern aircraft should have been invented earlier.

    B. Modern aircraft have powerful engines and can travel very fast.

    C. Aircraft are more useful than cars.

D. Someday cars will be replaced by small planes.

EXTENSIVE READING

Reading Decimal Fractions

The chart shown in Figure 1 – 12 gives the names of the parts of a number with respect to their positions from the decimal point.

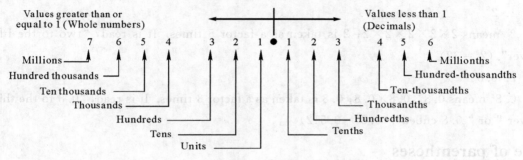

Figure 1 – 12

To read a mixed decimal (a whole number and a decimal fraction), read the whole number, read the word and at the decimal point, and read the decimal.

Examples

1. 2. 65 is read, "two and sixty-five hundredths".

2. 9. 002 is read, "nine and two thousandths".

3. 135. 078, 7 is read, "one hundred thirty-five and seven hundred eighty-seven ten thousandths".

Simplified Method of Reading Decimal Fractions

Often a simplified method of reading decimal fractions is used in actual on-the-job applications. This method is generally quicker, easier, and less likely to be misinterpreted. A tool and die maker reads 0. 018, 7 inch as "point zero, one eight, seven inches". An electronics technician reads 2. 125 amperes as "two, point one, two, five amperes".

Powers and Roots of Decimal Fractions

Powers of numbers are used to find the area of square surfaces and circular sections. Volumes of cubes, cylinders, and cones are determined by applying the power operation. Determining roots of numbers is used to find the lengths of sides and heights of certain geometric figures. Both powers and roots are required operations in solving many formulas in the electrical, machine, construction, and business occupations.

Meaning of Powers

Two or more numbers multiplied to produce a given number are factors of the given number. Two factors of 8 are 2 and 4. The factors of 15 are 3 and 5. A power is the product of two or more equal factors. An exponent shows how many times a number is taken as a factor. It is written smaller than the number, above, and to the right of the number.

Examples:

Find the indicated powers for each of the following.

1. 2^5

2^5 means $2 \times 2 \times 2 \times 2 \times 2$; 2 is taken as a factor 5 times. It is read, "two to the fifth power". ($2^5 = 32$)

2. 0.8^3

0.8^3 means $0.8 \times 0.8 \times 0.8$; 0.8 is taken as a factor 3 times. It is read, "0.8 to the third power" or "0.8 cubed". ($0.8^3 = 0.512$)

Use of parentheses

Parentheses are used as a grouping symbol. When an expression consisting of operations within parentheses is raised to a power, the operations within the parentheses are performed first. The result is then raised to the indicated power.

Example:

Raise to the indicated power. $(1.4 \times 0.3)^2$

Perform the operations within the parentheses first. $(1.4 \times 0.3)^2 = 0.42^2 = 0.176,4$

Raised to the given power.

Parentheses that enclose a fraction indicate that both the numerator and denominator are raised to the given power.

$$(3/4)^2 = 3/4^2 = 9/16 = 0.562,5$$

The same answer is obtained by dividing first and squaring second, as by squaring both terms first and dividing second.

$$(3/4)^2 = 0.75^2 = 0.562,5$$

Speed of Sound

In any uniform medium, under given physical conditions, sound travels at a definite speed. However, in some substances, the velocity of sound is higher than in others. As a general rule, the denser the medium, the faster sound travels. This is why sound travels much faster in water than in air. The speed of sound varies directly with the temperature of the air.

Mach Number

When studying aircraft that fly at supersonic speeds, it is customary to represent aircraft speed in relation to the speed of sound. The term Mach number is used to represent the ratio of the speed of the airplane to the speed of sound in the same atmospheric conditions.

Resonance

The natural frequency of a given object is the frequency where that object vibrates naturally, or without an external force. If two objects have the same natural frequency and are set next to each other, when one of them vibrates, it can transfer its wave energy to the other object making it vibrate. This transfer of energy is known as resonance. Because resonance induces vibration, it can exert destructive forces on an aircraft. For example, it is possible to have portions of an aircraft, such as the propeller, vibrate in resonance at certain engine speeds. If the vibrations are strong enough they can create stresses in the aircraft that could lead to structural failure.

Lesson 2　Hand Tools and Measuring Tools

INTENSIVE READING

Aviation Maintenance Hand Tools

The term "hand tools" encompasses all of the hand tools most commonly used in everyday maintenance and repair. Some of these tools are common, while others have a very specialized usage. Regardless of how common the tool, the subtleties for their proper use are not always known. Therefore, the most common uses are listed in the following text. Hand tools will fall into four basic categories: pounding tools, cutting tools, turning tools, and holding tools.

Pounding tools and cutting tools are more often used in shop maintenance such as making sheet metal parts, while turning tools and holding tools are more often used in line maintenance to install or remove the aircraft components.

Pounding Tools

Pounding tools include different types and weights of hammers and mallets, each with a very specific use. Since misuse of pounding tools can result in damage to aircraft components and injury to personnel, it is important that you always choose the one best suited for the job and use these tools properly.

Hammers

Figure 2 – 1 shows some of the hammers that the aviation mechanic may be required to use.

Before using a hammer, you should make sure the handle is secure and in good condition. When using any hammer or mallet, always take advantage of all the mechanical force available to you, swing the hammer from the elbow, not the wrist, and hold the hammer as far out on the end of the handle as possible while maintaining a firm grip. Always strike the work squarely using the full face of the hammer. To prevent marring the work, keep the face of a hammer or mallet smooth and free of dents.

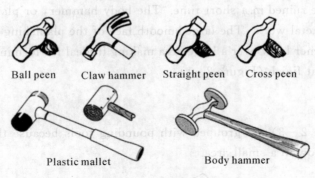

Ball peen   Claw hammer   Straight peen   Cross peen

Plastic mallet   Body hammer

Figure 2 - 1   Hammers

Metal-head hammers are usually sized according to the weight of the head without the handle.

Ball peen hammers range in weight from about one ounce to about three pounds. One hammer face is always flat while the other is formed into the shape of a ball. The flat hammer face is used for pounding on hard steel, but should not be used to drive a nail, since the curved face of a claw hammer is better shaped for nail driving control. The ball end of the hammer is typically used to peen over rivets in commercial sheet metal work. However, this is not the method used for securing rivets in aircraft sheet metal work (Figure 2 - 2). A claw hammer is a tool primarily used for pounding nails into, or extracting nails from, some other object. The head of a claw hammer is typically hardened, making it more brittle and susceptible to chipping. Therefore, a claw hammer should not be used on hardened steel parts. Commercial sheet metal often requires metal to be bent by hammering. This is typically accomplished by using either a cross peen or straight peen hammer.

Unlike the ball peen hammer, the cross and straight peen hammers have a wedge-type end that is used to either crease metal to start a bend, or to straighten out a rolled edge (Figure 2 - 1).

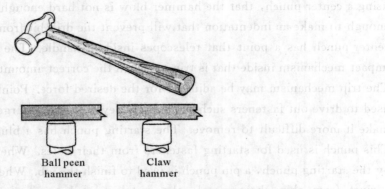

Ball peen   Claw
hammer   hammer

Figure 2 - 2   A ball peen hammer with a flat face and a ball-shaped end

The soft faced hammer, or mallet, is used for forming and shaping soft aluminum. Plastic mallets have replaceable tips and should never be used to drive nails or punches. The plastic head would be ruined in a short time. The body hammer, or planishing hammer, is also used for sheet metal work. The large smooth face of the planishing hammer distributes the force of the hammer blow over a large area making it ideal for forming and shaping sheet metal to a smooth flat finish (Figure 2 - 1).

Punches

Punches (Figure 2 - 3) are grouped with pounding tools because they concentrate the force of a hammer blow in a small area.

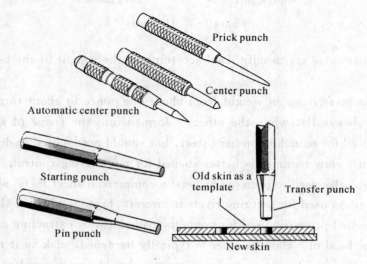

Figure 2 - 3   Punches

The prick punch has a sharp point, and is used to mark dimensions and locations on sheet metal. The center punch also has a sharp point, is usually more massive than the prick punch and is ground to a much shallower angle. The center punch is used to make an indentation in sheet metal as an alignment mark for start a drill. Care must be taken when using a center punch, that the hammer blow is not hard enough to distort the metal, but is enough to make an indentation that will prevent the drill bit from wandering. The automatic center punch has a point that telescopes inside a handle. The handle has a spring loaded impact mechanism inside that is tripped when the correct amount of pushing force is applied. The trip mechanism may be adjusted for the desired force. Pointed punches should never be used to drive out fasteners such as rivets. They will only enlarge the end of the fastener and make it more difficult to remove. The starting punch has a blunt tip and a tapered shank. This punch is used for starting fasteners from their holes. When the holes are nearly filled by the starting punch, a pin punch is used to finish the job. When using a pin punch to drive a rivet from thin sheet metal, the metal must be supported from behind to prevent distortion. The transfer punch is usually about 4 inches long. It has a point that tapers, then turns straight for a short distance in order to fit a drill locating hole in a template. The tip has a point similar to that of a prick punch. As its name implies, the transfer punch is used to transfer the location of holes through the template or pattern to the material.

Cutting Tools

The cutting tools used in aviation maintenance go far beyond the snips and saws that probably come to mind. In fact, cutting tools include any tool that removes or separates material.

Chisels

Chisels (Figure 2 – 4) are the most common and simplest of the cutting tools.

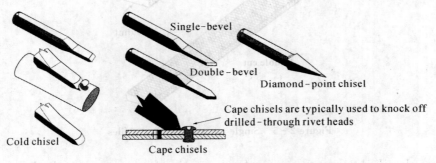

Figure 2 – 4   Chisels

They are usually made of high grade tool steel and are heat treated and tempered for maximum performance on a variety of materials. The cold chisel is used for cutting metal and has its cutting edge ground to an angle of about 70°. The chisel is ground slightly convex. With this design, most of the force of the hammer blow is directed to the center of the cutting edge. This convex shape also holds the corners of the chisel away from the work. This helps prevent nicking the metal near the actual cutting point. The cape chisels come in either single-bevel, or double-bevel. The cape chisel has a much narrower cutting edge than the cold chisel and is used to cut channels or key ways. The cape chisel is also the preferred tool for cutting the heads of rivets. The narrow cutting area helps prevent extraneous metal damage to the area around the rivet. A diamond-point chisel is forged into a sharp-cornered square section and then ground to an acute angle. This forms a cutting edge that is similar in shape to a diamond. These chisels are used to cut V-grooves and sharp corners in square or rectangular grooves.

Files

It is impossible to do quality sheet metal work without the aid of a file. This cutting tool has rows of teeth shaped like tiny chisels cut diagonally across its face. A file with all of its teeth cut in the same direction is a single cut file, and is used to produce a smooth finish on the material being cut. The double cut file has two sets of teeth cut at angles to each other, and is designed to cut more metal with each path than the single cut file (Figure 2 – 5).

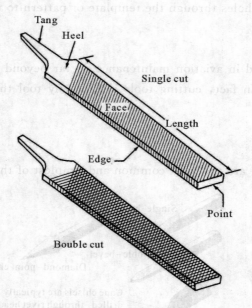

Figure 2 – 5   Single cut and double cut files

Files are classified as to coarseness of cut, ranging from coarse, to dead smooth.

Different types of file are designed for different jobs (Figure 2 – 6). Be sure you select the proper file for the work at hand. When using the file, pressure should be applied on the forwards stroke only, drawing the file backward while maintaining pressure will dull or damage the teeth. Each stroke should use the full length of the file, and the pressure should be even throughout the stroke. Each stroke should be smooth and slow. The material should be held firmly in a vise or clamp to prevent it from chattering. Files should be cleaned often during use. Metal chips left in the teeth can score and damage the work. A wire brush called a file card and a metal pick are designed for the job. Files are made of high carbon steel which makes them fairly brittle. They should be covered and stored where they will not come in contact with each other or with other metal tools. This will help prevent breakage of the teeth. Never oil files, always store them dry and in a dry place. The high carbon steel makes file susceptible to rust. A rusty file is worthless as cutting tool.

Hand snips

Shears are another type of cutting tool used on aircraft sheet metal. Tin snips are basically used for making straight cuts; they are also used to cut curves to either the left or right (Figure 2 – 7).

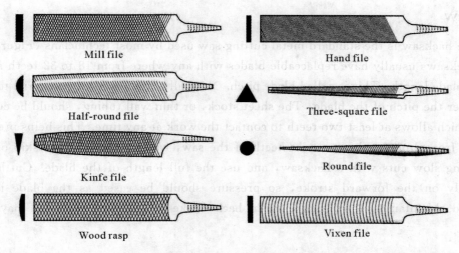

Figure 2 – 6   Files are classified with regard to their kind, their length, and their cut

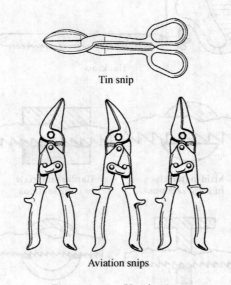

Figure 2 – 7   Hand snips

The aviation maintenance technician will occasionally fabricate sheet metal parts requiring cutting and trimming. Special aviation snips designed with serrated blades and compound leverage make such work fairly simple. Aviation snips are designed to make straight cuts for cutting to the right and cutting to the left. The straight snips are color coded with a yellow handle, the right-hand snips have a green handle, and those that cut to the left have a red handle. In addition to being color-coded, aviation snips can be identified by their shape. For example, a straight snip has a relatively straight nose. However, with right-cut snips, when held in your hand, the lower jaw is on the right whereas the lower jaw is on the left with left-cut snips.

Hacksaw

The hacksaw is the standard metal cutting saw used by most technicians (Figure 2 – 8).

Hacksaws usually have replaceable blades with any where from 18 to 32 teeth for every inch of blade length. This is called blade pitch. Normally, the harder the material to be cut, the higher the pitch of the blade. The sheet stock, or thin wall tubing, should be cut with a blade which allows at least two teeth to contact the work at any time. This helps prevent the material from dropping between the teeth of the saw, causing the blade to skip or break. Make long slow cuts with a hacksaw, and use the full length of the blade. Cutting takes place only on the forward stroke, so pressure should be eased as the blade is drawn backward. Maintaining the pressure on the back stroke will dull the blade and may cause it to break.

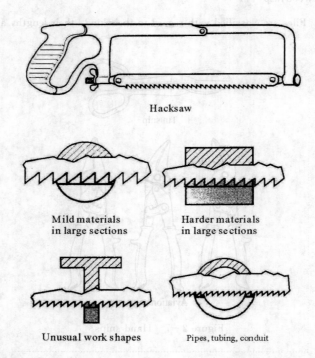

Hacksaw

Mild materials in large sections

Harder materials in large sections

Unusual work shapes

Pipes, tubing, conduit

Figure 2 – 8   Typical uses for various pitch hacksaw blades

Twist drills

The most important tool to any one who does any amount of sheet metal work on an aircraft is the twist drill (Figure 2 – 9). The twist drill is inserted into the chuck of a drill motor and secured with a chuck key. The chuck grips the drill by the shank. The body of the drill is cut with spiral flutes that help carry metal chips away from the point and allow lubrication to reach the point. The land between the flutes is ground so that only a small portion called the margin is actually in contact with the side of the hole during the drilling operation. The remainder of the land is smaller than the hole to help prevent the drill from bending. For drilling most aircraft aluminum, the point of the drill is ground to 50°on either side of the center line for an included angle of 118°. The lip relief or heel angle is ground back from the cutting edge by about 12°- 15°. Too flat an angle here would not allow the cutting edge to cut. It would act the same as a knife blade placed flat on the surface to be cut. No cutting can take place until the blade is given some angle and allow to dig into the material. Always wear eye protection when drilling, and check the recommended drill speed for the material being cut. Higher speeds are required for drilling soft materials while the lower speeds are used for hard metals such as stainless steel. Cutting oil should be used when drilling steel while no lubrication is generally required for drilling aluminum.

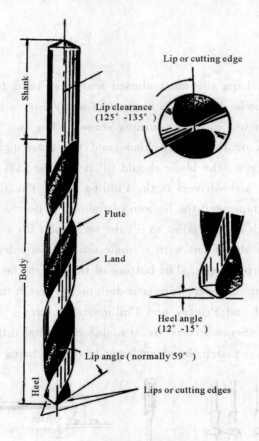

Figure 2 - 9 Twist drill

Reamer

Where tolerances are critical, holes are drilled slightly smaller than the fastener's being used, then are finished to exact dimensions with a reamer. The reamer has precision ground cutting blades that remove small amount of metal as it is turned into the hole. Reamers are always turned in the cutting direction both entering and leaving the hole. Once the hole has been reamed to exact dimensions, the fastener is pressed or tapped into position (Figure 2 – 10).

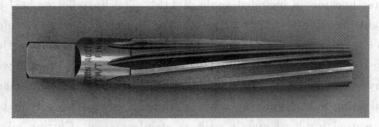

Figure 2 – 10   Morse taper reamer

Turning Tools

Screwdriver

The screwdriver is perhaps the most abused and miss used tool in the mechanics' toolbox. Although its shape lends itself to be used as a pry bar, a punch, or a chisel, the screw driver is designed for one purpose: turning screws. The two most common types of screwdrivers are the plain or the slotted blade and the cross point. When choosing a screwdriver for a slotted screw, the blade should fill it at least 75% of the slot. The most common of the cross point screwdrivers is the Phillips head. The slot in the Phillips head screw is cut with a double taper and the bottom of the slot is nearly flat. The Phillips head screwdriver is designed with a blunt point to fill the screw slot for maximum turning force. The Reed and Prince screw slot is cut with a single taper. The sides of the slot are nearly parallel and form a near perfect cross. The bottom of the slot in the Reed and Prince screw comes to a fairly sharp point. The screwdriver is designed to match this point. Interchanging screwdrivers between Reed and Prince, and Phillips head screws could damage both the screw and the screwdriver. Screwdrivers are available in several different types and sizes. The technician should always match the screwdriver to the job being done (Figure 2 – 11).

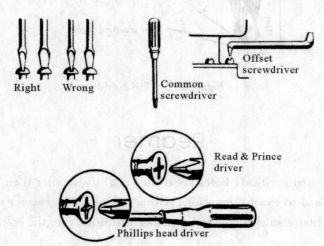

Figure 2 – 11   Screw drivers and their uses

Wrenches

Most of the threaded fasteners on modern aircraft require wrenches to turn them. The assortment of wrenches available to the aviation maintenance technician is almost endless, but a few basic types are essential.

The open-end wrench has an opening with parallel sides, and is designed to accommodate specific sizes of nuts and bolt heads. The opening is usually off-set at an angle of about 15°. This allows the wrench to be turned over to fit the fastener in spaces where room for the handle is limited (Figure 2 – 12).

The box-end wrench is used mainly for breaking fasteners loose. The box-end wrench has a six or twelve point opening which is off-set from the axis of the handle by about 15°. The twelve point wrench may be repositioned on the nut or bolt head every 30° while there must be at least 60°of swing space in order to reposition the six point wrench (Figure 2 – 12).

The combination wrench incorporates an open end wrench on one end and the same size box end wrench on the other (Figure 2 – 12).

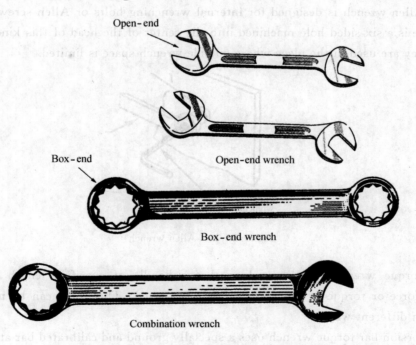

Open-end

Box-end

Open-end wrench

Box-end wrench

Combination wrench

Figure 2 – 12   Open-end, box-end and combination wrenches

The socket wrench is one of the most versatile tools in the mechanic's tool box. It has a six or twelve point opening on one end, and a square opening on the opposite end to accommodate any of several different handles, including the speed handle, the breaker bar, the nut driver, and the ratchet handle. These handles may be used with various extensions and universal joints to reach fasteners located in awkward positions. The ratchet handle is a special useful because it may be turned and repositioned without removing the wrench from the fastener. The ratchet locks the handle so it will turn in one direction but will slip or ratchet in the other direction. It may be adjusted for turning or ratcheting in either direction. The socket may also be used with a special handle designed for measuring the twisting force used when tightening a fastener (Figure 2 – 13).

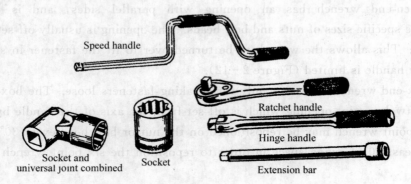

Figure 2 – 13   Socket wrench set

The Allen wrench is designed for internal wrenching bolts or Allen screw (Figure 2 – 14). There is a six-sided hole machined into the center of the head of this kind of bolt and screw. They are used in the place where outside wrench space is limited.

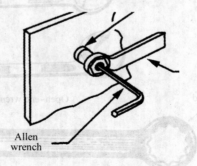

Figure 2 – 14   Allen wrench

The torque wrench is calibrated so that a handle of specified length will apply a measured force or torque to the fastener as it is turned. This force can be measured and indicated in different ways.

The torsion bar torque wrench uses a specially ground and calibrated bar attached to the lug, and to a gear system. As the wrench is turned, the torsion bar inside the wrench is twisted, the amount of twist is translated into a reading on a dial through the gear system. Torsion bar torque wrenches requires that the technician be in a position to see the torque readings on the wrench.

The toggle type torque wrench: there are times, the location of a fastener will not allow the technician a clear view of the indicator. To solve this problem, the toggle type torque wrench may be the answer[4]. The desired torque value is set into the wrench by turning the handle. This compresses a spring inside the handle. When the wrench is turned until the desired torque is reached, a unique toggle mechanism in the handle overcomes the spring tension, torque is released momentarily to allow a few degrees of handle rotation. The operator feels the release as a click or thump in the handle. This gives the technician a positive indication that the preset torque value has been reached but does not require a visual display.

Extensions may be used with torque wrenches to reach awkward locations, but on the certain circumstances, use of an extension will change the torque value of the wrench. Text books and Maintenance Manuals usually include the proper formula for determining the accurate torque value when an extension is used (Figure 2-15). As with any precision tool, the accuracy of the torque wrench should be checked and recalibrated at regular intervals.

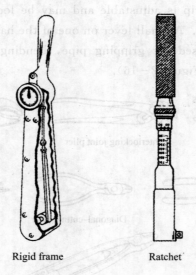

Rigid frame   Ratchet

Basic formula $FL = T$

F—Applied force;

L—Lever length betwen centerline of drive and centerline of applied force (F must be 90 degrees to L);

T—Torque.

Formula for use with extensions $T_w = \dfrac{T_e A}{B}$

A—Lever length of wrench;

B—Lever length of wrench plus extension;

T_e—Required torque on bolt;

T_w—Torque reading on wrench dial.

Figure 2-15   Common types of torque wrench

Holding Tools

Pliers

Pliers are perhaps the most familiar of the holding tools, used in the home and industry.

Slip joint pliers: the six inch slip joint pliers are probably the most popular of the holding tools. The pivot point between the jaws may be moved into one of two different holes in one of the jaws. This provides a wide range of grip sizes (Figure 2 – 16).

Interlocking joint pliers are also adjustable for various grip sizes. Here, one jaw may be slipped from one channel to another in the opposite jaw. Sometimes called water pump pliers, they are also known by the trade name, channel locks (Figure 2 – 16).

Vise grips: another familiar trade name is given to a special set of high leverage pliers called vise grips. The vise grip is adjustable and may be locked into place by a specially designed toggle in the handles. A small lever on one of the handle is used to unlock the vise grip. These tools may be used for gripping pipe, bending metal, pinching tubing, or clamping metal for welding (Figure 2 – 16).

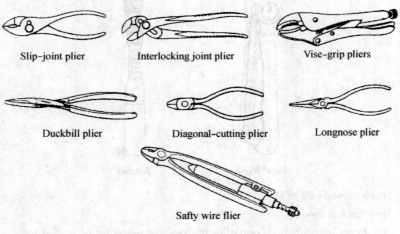

Slip-joint plier   Interlocking joint plier   Vise-grip pliers

Duckbill plier   Diagonal-cutting plier   Longnose plier

Safty wire flier

Figure 2 – 16   Types of pliers

Duckbill pliers are valuable tools to the aviation maintenance technician. The long handles and flat jaws are well suited for twisting safety wires. The serrations in the jaws of the duckbills are not as deep or sharp as in many other pliers. This feature helps prevent nicking and weakening of the safety wire as it being installed (Figure 2 – 16).

Diagonal pliers are usually referred to as diagonals. The diagonal is a short-jawed cutter with a blade set at a slight angle on each jaw. This tool can be used to cut wire, rivets, small screws, and cotter pins, besides being practically in removing or installing safety wire (Figure 2 – 16).

The duckbill pliers and the diagonal cutting pliers are used extensively in aviation for the job of safety wiring.

Long nose or needle nose pliers are a special useful in tight quarters. They are also helpful in gripping small parts and are ideal for electrical soldering work. These versatile tools are found in several different styles to perform a variety of jobs (Figure 2 – 16).

Safety wire pliers: special safety wire pliers have been designed specifically for the aviation industry. The jaws are flat with slight serrations. The handles are equipped with twisting mechanism. The safety wire is passed through the fastener, for instance, a bolt head. The two ends are brought together and trimmed to the proper length with the built-in cutter. The jaws are locked in place over the ends of the wire. Pulling the handle creates a twisting motion that wraps the two wires around each other. When the handles are unlocked, the safety wiring is completed, and the ends may be trimmed (Figure 2 – 16).

New Words/ Phrases/ Expression

1. encompass[in'kʌmpəs]   v. 包含
2. subtlety ['sʌtlti]   n. 精细，细微的差别
3. turning tool   拧动工具
4. holding tool   夹持工具
5. pounding tool   敲击工具
6. cutting tool   切割工具
7. line maintenance   航线维修
8. shop[ʃɒp]   n. 车间
9. sheet metal   n. 钣金工
10. hammer['hæmər]   n. 锤子，榔头
11. ball hammer   圆头锤
12. claw hammer   羊角锤
13. straight hammer   直头锤
14. cross hammer   横头锤
15. plastic hammer   塑料锤
16. body hammer   打平锤，锤光锤
17. prick punch   划线冲子，冲孔器中心冲头
18. center punch   中心冲
19. automatic punch   自动中心冲
20. starting punch   起始冲
21. pin punch   销冲，销冲头
22. transfer punch   转换冲
23. cold chisel   冷錾，冷凿
24. cape chisel   前扁尖凿，斜刃凿
25. diamond point chisel   棱形凿，金刚石尖头凿
26. V-groove   V 形槽

27. mill file　扁锉
28. hand file　手锉,平锉
29. half round file　半圆锉
30. three-square file　三角锉
31. knife file　刀型锉,刀锉
32. round file　圆锉
33. wood rasp　木锉
34. vixen file　弧纹锉,弯纹锉
35. tin snip　铁皮剪
36. aviation snip　航空剪
37. hacksaw['hæk,sɔ:]　n.钢锯,弓锯
38. twist drill　麻花钻
39. chuck[tʃʌk]　n.卡盘
40. spiral flute　螺槽
41. reamer　['ri:mə]　n.铰刀
42. abuse[ə'bju:z]　v.& n.滥用
43. taper['teipə]　n.锥度
44. parallel['pærəlel]　adj.& n.平行的,平行线
45. interchange[,intə'tʃeindʒ]　v.& n.互换
46. offset['ɔf,set]　v.& n.偏置
47. pry bar[prai][ba:]　n.撬杆,撬杠
48. punch[pʌntʃ]　n.冲子
49. chisel['tʃizl]　n.凿子
50. screw driver　n.螺丝刀
51. slotted screw　n.带槽螺钉(一字螺钉)
52. cross point screw driver　n.十字螺刀
53. blunt[blʌnt]　adj.钝的
54. wrench[rentʃ]　n.扳手
55. open-end wrench　n.开口扳手
56. solid wrench　n.呆扳手(开口扳手的别称)
57. box-end wrench　n.梅花扳手
58. socket wrench　n.套筒扳手
59. Allen wrench　n.内六角扳手
60. Allen screw　n.内六角螺钉
61. accommodate[ə'kɔmədeit]　v.容纳
62. speed handle　n.快速摇把
63. ratchet['rætʃit]　n.棘轮,棘齿
64. awkward['ɔ:kwəd]　adj.别扭的,难操纵的
65. calibrate['kæli,breit]　v.校准

66. torsion bar torque wrench   *n.* 扭力杆式力矩扳手

67. desire[di'zaiə]   *v.* & *n.* 渴望

68. visual['vizjuəl]   *adj.* 目视的,视力的

69. circumstance['sə:kəmstəns]   *n.* 情况,环境,条件

70. torque value   *n.* 力矩值

71. accurate['ækjurit]   *adj.* 精确的

72. accuracy['ækjurəsi]   *n.* 精确性,准确性

73. interval['intəvəl]   *n.* 间隔时间

74. pliers['plaiəz]   *n.* 钳子

75. slip joint pliers   鱼口钳

76. interlocking joint pliers   内锁支点钳

77. leverage['levə,ridʒ]   *n.* 杠杆作用,杠杆效率

78. vise grip   大力钳

79. toggle['tɔgl]   *n.* 开关,触发器

80. duckbill pliers   *n.* 鸭嘴钳

81. nick[nik]   *n.* 刻痕;凹隙

82. needle nose pliers   尖嘴钳

83. versatile['və:sətail]   *adj.* 万向的;万能的

84. jaw[dʒɔ:]   *n.* 颚,钳口

85. clamp[klæmp]   *v.* & *n.* 夹紧,夹钳

86. serration[se'reiʃən]   *n.* 锯齿

87. diagonal[dai'ægənl]   *adj.* 对角线的,斜的

88. solder['sɔldə]   *v.* & *n.* 焊接,锡焊

89. trim[trim]   *v.* 修整,配平

90. wrap[ræp]   *v.* 卷,裹

Notes

(1) Since misuse of pounding tools can result in damage to aircraft components and injury to personnel, it is important that you always choose the one best suited for the job and use these tools properly.

分析:misuse 的含义为"误用"。整个句子前半部分表示原因,后半部分表示结果。并且,后半部分为"it is ... that ..."的句式结构,it 为形式主语,that 后面的句子为真正的主语。

翻译:由于敲击工具的误用会导致飞机部件的损伤和人员的伤亡,所以,时常为你的工作选择最合适的工具和正确的使用它具有非常重要的意义。

(2) The cutting tools used in aviation maintenance go far beyond the snips and saws that probably come to mind.

分析:"The cutting tools"为主语,"used in aviation maintenance"为主语的定语,"go far beyond"为整个句子的谓语,"the snips and saws"为宾语成分,而"that probably come to mind"为宾语的定语成分。

翻译:在航空维修中用到的切割工具(的种类)远远超过(我们)能想象到的剪刀和锯子。

(3)The Allen wrench is designed for internal wrenching bolts or Allen screw. There is a six-sided hole machined into the center of the head of this kind of bolt and screw. They are used in the place where outside wrench space is limited.

分析:Allen wrench 直译艾伦扳手,实际是指内六角扳手。

翻译:内六角扳手是为内六角螺栓或内六角螺钉设计的,在这类螺栓和螺钉的头部中央,开有六边形的孔。这类螺栓和螺钉用于(螺栓头或螺钉头)外部的扳手空间很有限的地方。

(4)To solve this problem, the toggle type torque wrench may be the answer.

分析:the toggle type torque wrench 应按中国对此类工具的习惯称谓翻译为"定力矩扳手"更为准确易懂。

翻译:为了解决这一问题,可用触发式力矩扳手(定力矩扳手)。

Exercises

Ⅰ. **Answer the following questions.**

(1) Why should the plastic mallets never be used to drive nails or punches?

(2) Which condition is best for planishing hammer?

(3) Why Pointed punches should never be used to drive out fasteners such as rivets?

(4) Is it right to draw the files backward while maintaining pressure?

(5) How to choose a hacksaw to cut a sheet stock?

(6) What function the convex shape of chisel is?

(7) List some hand tools before you view the text.

(8) Why the Phillips headscrew driver is designed with a blunt point?

(9) List some wrenches that can measure and indicate the force.

(10) Which tool may be turned and repositioned without removing the wrench from the fastener?

(11) How does the slip joint pliers to provide a wide range of grip size?

(12) What kinds of jobs could the needle nose pliers used for?

Ⅱ. **Translating the following sentences into Chinese.**

(1) When using a pin punch to drive a rivet from thin sheet metal, the metal must be supported from behind to prevent distortion.

(2) Cutting takes place only on the forward stroke, so pressure should be eased as the blade is drawn backward. Maintaining the pressure on the back stroke will dull the blade and may cause it to break.

(3) Higher speeds are required for drilling soft materials while the lower speeds are used for hard metals such as stainless steel.

(4) Hand tools will fall into four basic categories: turning tools, holding tools,

pounding tools and cutting tools.

(5) Although its shape lends itself to be used as a pry bar, a punch, or a chisel, the screw driver is designed for one purpose: Turning screws.

(6) Interchanging screwdrivers between Reed and Prince, and Phillips head screws could damage both the screw and the screw driver.

(7) The torque wrench is calibrated so that a handle of specified length will apply a measured force or torque to the fastener as it is turned.

(8) The socket wrench has a six or twelve point opening on one end, and a square opening on the opposite end to accommodate any of several different handles.

(9) The torsion bar torque wrench uses a specially ground and calibrated bar attached to the lug,and to a gear system.

(10) The pivot point between the jaws may be moved into one of two different holes in one of the jaws. This provides a wide range of grip sizes.

Ⅲ. **Choose the right English meaning for following terms.**

(1) 夹持工具

　　A. holding tool　　　　　　B. turning tool　　　　　　C. cutting tool

(2) 螺刀

　　A. screwdriver　　　　　　B. wrench　　　　　　　　C. extension

(3) 冲头、冲子

　　A. chisel　　　　　　　　B. punch　　　　　　　　C. hammer

(4) 偏置螺刀

　　A. common screwdriver　　B. Philips head driver　　C. offset screwdriver

(5) 套筒扳手

　　A. socket wrench　　　　　B. box-end wrench　　　　C. the Allen wrench

(6) 鱼口钳,鲤鱼钳

　　A. slip-joint plier　　　　B. vise-gripplier　　　　C. diagonal-cutting plier

(7) 滥用

　　A. misuse　　　　　　　　B. abuse　　　　　　　　C. awkward

(8) 圆头锤

　　A. cross peen hammer　　　B. ball peen hammer　　　C. claw hammer

(9) 凿子

　　A. chisel　　　　　　　　B. punch　　　　　　　　C. file

Ⅳ. **choose the right English meaning for following terms.**

(1) Which hammer can be used for nail driving control?

　　A. ball peen hammer　　　　　　　B. claw hammer

　　C. straight peen hammer　　　　　　D. cross peen hammer

(2) Which hammer can be used to peen over rivets in aircraft sheet metal work?

　　A. ball peen hammer　　　　　　　B. claw hammer

C. straight peen hammer   D. cross peen hammer

(3) Which kind of the hammers has a wedge-type end that is used to either crease metal to start a bend, or to straighten out a rolled edge?

    A. ball peen hammer   B. claw hammer

    C. plastic mallet   D. cross peen hammer

(4) Which kind of the hammers is be used for sheet metal work.

    A. ball peen hammer   B. the body hamer

    C. claw hammer   D. cross peen hammer

(5) Which kind of the punches has a sharp point, is used to mark dimensions and locations on sheet metal?

    A. pin punch   B. starting punch   C. center punch   D. prick punch

(6) Which kind of the punches is used to make an indentation in sheet metal as an alignment mark for start a drill?

    A. prick punch   B. center punch   C. starting punch   D. pin punch

(7) Which kind of the puncher has a point that telescopes inside a handle?

    A. prick punch   B. automatic center punch

    C. starting punch   D. pin punch

(8) Which kind of the puncher is used to transfer the location of the holes through the template or pattern to the material?

    A. pin punch   B. starting punch   C. center punch   D. transfer punch

(9) Which of the four following cutting tools are the most common and simplest of the cutting tools?

    A. hacksaw   B. chisels   C. hand snips   D. files

(10) The cold chisel is used for cutting metal and has its cutting edge ground to an angle of (   ).

    A. 75°   B. 60°   C. 65°   D. 70°

EXTENSIVE READING

Aviation Maintenance Measuring Tools

Measuring devices are precision tools. They are carefully machined, accurately marked and, in many cases, are made up of very delicate parts. When using these tools, be careful not to drop, bend, or scratch them. The finished product will be no more accurate than the measurements or the layout; therefore, it is very important to understand how to read, use, and care for these tools.

Combination Sets

The combination set (Figure 2 – 17), as its name implies, is a tool that has several uses. It can be used for the same purposes as an ordinary tri-square, but it differs from the tri-square in that the head slides along the blade and can be clamped at any desired place. Combined with the square or stock head are a level and scriber. The head slides in a central groove on the blade or scale, which can be used separately as a rule. The spirit level in the stock head makes it convenient to square a piece of material with a surface and at the same time tell whether one or the other is plumb or level. The head can be used alone as a simple level. The combination of square head and blade can also be used as a marking gage to scribe lines at a 45° angle, as a depth gage, or as a height gage. A convenient scriber is held frictionally in the head by a small brass bushing. The center head is used to find the center of shafts or other cylindrical work. The protractor head can be used to check angles and also may be set at any desired angle to draw lines.

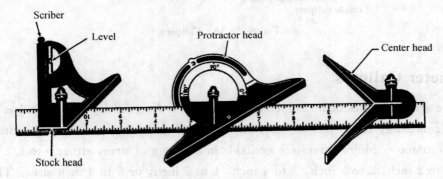

Figure 2 – 17　Combination set

Dividers and Pencil Compasses

Dividers and pencil compasses have two legs joined at the top by a pivot. They are used to scribe circles and arcs and for transferring measurements from the rule to the work. Pencil compasses have one leg tapered to a needle point; the other leg has a pencil or pencil lead inserted. Dividers have both legs tapered to needle points.

Calipers

Calipers are used for measuring diameters and distances or for comparing distances and sizes. The three common types of calipers are the inside, the outside, and the hermaphrodite calipers, such as gear-tool calipers (Figure 2 – 18).

Outside calipers are used for measuring outside dimensions, for example, the diameter of a piece of round stock. Inside calipers have outward curved legs for measuring inside diameters, such as diameters of holes, the distance between two surfaces, the width of slots, and other similar jobs. A hermaphrodite caliper is generally used as a marking gage in layout work. It should not be used for precision measurement.

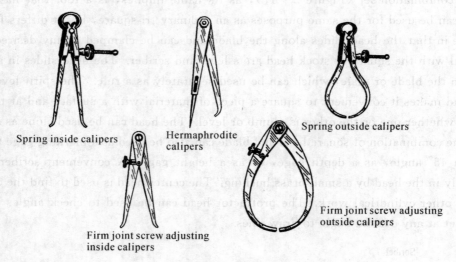

Spring inside calipers

Hermaphrodite calipers

Spring outside calipers

Firm joint screw adjusting outside calipers

Firm joint screw adjusting inside calipers

Figure 2 – 18   Calipers

Micrometer Calipers

There are four types of micrometer calipers, each designed for a specific use. The four types are commonly called outside micrometer, inside micrometer, depth micrometer, and thread micrometer. Micrometers are available in a variety of sizes, either 0 to 1/2 inch, 0 to 1 inch, 1 to 2 inch, 2 to 3 inch, 3 to 4 inch, 4 to 5 inch, or 5 to 6 inch sizes. The outside micrometer (Figure 2 – 19) is used by the mechanic more often than any other type. It may be used to measure the outside dimensions of shafts, thickness of sheet metal stock, diameter of drills, and for many other applications.

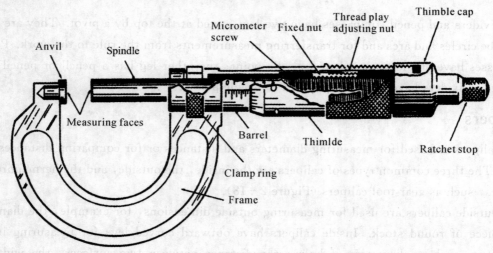

Thimble cap

Thread play adjusting nut

Micrometer screw

Fixed nut

Anvil

Spindle

Measuring faces

Barrel

Thimlde

Ratchet stop

Clamp ring

Frame

Figure 2 – 19   Outside micrometer

The smallest measurement which can be made with the use of the steel rule is one sixty-fourth of an inch in common fractions, and one one-hundredth of an inch in decimal fractions. To measure more closely than this (in thousandths and ten-thousandths of an inch), a micrometer is used. If a dimension given in a common fraction is to be measured with the micrometer, the fraction must be converted to its decimal equivalent. All four types of micrometers are read in the same way. The method of reading an outside micrometer is discussed later in this chapter.

Micrometer Parts

The fixed parts of a micrometer (Figure 2 - 19) are the frame, barrel, and anvil. The movable parts of a micrometer are the thimble and spindle. The thimble rotates the spindle which moves in the threaded portion inside the barrel. Turning the thimble provides an opening between the anvil and the end of the spindle where the work is measured. The size of the work is indicated by the graduations on the barrel and thimble.

Reading Micrometer

The lines on the barrel marked 1, 2, 3, 4, etc. , indicate measurements of tenths, or 0. 100 inch, 0. 200 inch, 0. 300 inch, 0. 400 inch, respectively (Figure 2 - 20). Each of the sections between the tenths divisions (between 1, 2, 3, 4, etc.) is divided into four parts of 0. 025 inch each. One complete revolution of the thimble (from zero on the thimble around to the same zero) moves it one of these divisions (0. 025 inch) along the barrel(Figure 2 - 20).

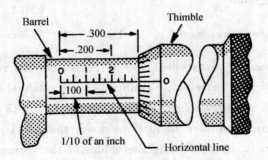

Figure 2 - 20   Micrometer measurements

The bevel edge of the thimble is divided into 25 equal parts. Each of these parts represents one twenty-fifth of the distance the thimble travels along the barrel in moving from one of the 0. 025-inch divisions to another. Thus, each division on the thimble represents one one-thousandth (0.001) of an inch. These divisions are marked for convenience at every five spaces by 0, 5, 10, 15, and 20. When 25 of these graduations have passed the horizontal line on the barrel, the spindle (having made one revolution) has moved 0. 025 inch.

The micrometer is read by first noting the last visible figure on tile horizontal line of the barrel representing tenths of an inch. Add to this the length of barrel between tile thimble and the previously noted number. (This is found by multiplying the number of graduations by 0.025 inch.) Add to this the number of divisions on the bevel edge of the thimble that coincides with the line of the graduation. The total of the three figures equals the measurement (Figure 2 – 21).

Vernier Scale

Some micrometers are equipped with a vernier scale which makes it possible to read directly the fraction of a division that may be indicated on the thimble scale. Typical examples of the vernier scale as it applies to the micrometer are shown in Figure 2 – 22. All three scales on a micrometer are not fully visible without turning the micrometer; but the examples shown in Figure 2 – 22 are drawn as though the barrel and thimble of tile micrometer were laid out flat so that all three scales can be seen at the same time. The barrel scale is the lower horizontal scale; tile thimble scale is vertical on the right; and the long horizontal lines (0 through 9 and 0) make up the vernier scale.

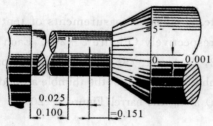

Figure 2 – 21　Reading a micrometer

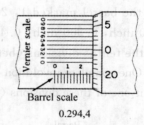

Figure 2 – 22　Vernier scale readings

In reading a micrometer, an excellent way to remember the relative scale values is to remember that the 0.025-inch barrel scale graduations are established by the lead screw (40 threads per inch). Next, the thimble graduations divide the 0.025 inch into 25 parts, each equal to 0.001 inch; then the vernier graduations divide the 0.001 inch into 10 equal parts, each equal to 0.000,1 inch. Remembering the values of the various scale graduations, the barrel scale reading is noted. The thimble scale reading is added to it; then the vernier scale reading is added to get the final reading. The vernier scale line to be read is always the one aligned exactly with any thimble graduation.

In figure 2 – 22, the barrel reads 0.275 inch and the thimble reads more than 0.019 inch. The number 1 graduation on the thimble is alined exactly with tile number 4 graduation on the vernier scale. Thus, the final reading is 0.294,4 inch.

Using a Micrometer

The micrometer must be handled carefully. If it is dropped, its accuracy may be permanently affected. Continually sliding work between the anvil and spindle may wear the surfaces. If the spindle is tightened too much, the frame may be sprung permanently and inaccurate readings will result.

To measure a piece of work with the micrometer, hold the frame of the micrometer in the palm of the hand with the little finger or third finger, whichever is more convenient. This allows the thumb and forefinger to be free to revolve the thimble for adjustment.

New Words/ Phrases/ Expression

1. delicate['delikit] *adj.* 精巧的，灵敏的，精密的
2. scratch *vt.* 乱涂，勾抹掉，擦，刮
3. accurate *adj.* 正确的，精确的
4. combination set 组合角尺（量具）
5. clamp *vt.* 夹住，夹紧
6. groove *n.* （唱片等的）凹槽
7. square *n.* 正方形，平方，直角尺
8. scriber *n.* 划线器，描绘标记的用具
9. dividers *n.* 两脚规，分规
10. compasses *n.* 圆规
11. pivot *n.* 枢轴，支点
12. needle *n.* 针
13. calipers *n.* 弯脚器，测径器，卡规
14. hermaphrodite *adj.* 具有相反性质的
15. micrometer *n.* 测微计，千分尺
16. frame *n.* 框架
17. barrel *n.* 圆筒
18. anvil *n.* （铁）砧，砧座
19. thimble *n.* 顶针，嵌环，套管
20. spindle *n.* 轴，杆，心轴
21. bevel *n.* 斜角，倾斜，斜面
22. graduation *n.* 毕业，刻度，分等级
23. multiply *v.* 繁殖，乘，增加
24. division *n.* 分开，区分，除法
25. coincide *vi.* 一致，符合
26. vernier *n.* [机]游尺，游标，游标卡尺

Exercises

Ⅰ. **Answer the following questions.**

(1) What function have the combination sets?

(2)How to distinguish between dividers and pencil compasses?

(3)Which type of calipers is used for measuring outside dimensions, for example, the diameter of a piece of round stock?

(4)Can a hermaphrodite caliper be used for precision measurement?

II. Put the following sentences into Chinese.

(1)The finished product will be no more accurate than the measurements or the layout; therefore, it is very important to understand how to read, use, and care for these tools.

(2)The spirit level in the stock head makes it convenient to square a piece of material with a surface and at the same time tell whether one or the other is plumb or level.

(3)The smallest measurement which can be made with the use of the steel rule is one sixty-fourth of an inch in common fractions, and one one-hundredth of an inch in decimal fractions.

(4)Some micrometers are equipped with a vernier scale which makes it possible to read directly the fraction of a division that may be indicated on the thimble scale.

Lesson 3   Aircraft Metals

INTENSIVE READING

Ferrous Aircraft Metals

Many different metals are required in the repair of aircraft. In selecting materials for aircraft repair, many factors are considered in relation to the mechanical and physical properties. Among the common materials used are ferrous metals. The term "ferrous" applies to the group of metals having iron as their principal constituent.

Identification

If carbon is added to iron, in percentages ranging up to approximately 1 percent, the product is vastly superior to iron alone and is classified as carbon steel. Carbon steel forms the base of those alloy steels produced by combining carbon steel with other elements known to improve the properties of steel. The addition of other metals changes or improves the chemical or physical properties of the base metal for a particular use.

Types, Characteristics, and Uses of Alloyed Steels

Plain Carbon Steel

Steel containing carbon in percentages ranging from 0. 10 to 0. 30 percent is classed as low-carbon steel. Steels of this grade are used for making such items as safety wire, certain nuts, cable bushings, or threaded rod ends.

Steel containing carbon in percentages ranging from 0. 30 to 0. 50 percent is classed as medium-carbon steel. This steel is especially adaptable for machining or forging, and where surface hardness is desirable.

Steel containing carbon in percentages ranging from 0. 50 to 1. 05 percent is classed as high-carbon steel. The addition of other elements in varying quantities adds to the hardness of this steel. In the fully heat-treated condition it is very hard, will withstand high shear and wear, and will have little deformation. It has limited use in aircraft.

Nickel Steel

The various nickel steels are produced by combining nickel with carbon steel. Nickel increases the hardness, tensile strength, and elastic limit of steel without appreciably decreasing the ductility.

Chromium Steel

Chromium steel is high in hardness, strength, and corrosion-resistant properties, and is particularly adaptable for heat-treated forgings which require greater toughness and strength than may be obtained in plain carbon steel. It can be used as the balls and rollers of antifriction bearings.

Stainless Steel：18 – 8 Steel

Stainless steel is used in many places on modern aircraft, such as fire walls, skins, structural parts, and special fasteners. Stainless steel is often called 18 – 8 steel. Because it contains 18% chrome alloyed to 8% nickel and 74% steel. Chrome forms an oxide which prevents corrosion from taking place.

Chrome-Vanadium Steel

The chrome-vanadium steels are made of approximately 18 percent vanadium and about 1 percent chromium. When heat treated, they have strength, toughness, and resistance to wear and fatigue. A special grade of this steel in sheet form can be cold-formed into intricate shapes. It can be folded and flattened without signs of breaking or failure.

Chrome-Molybdenum Steel

Molybdenum in small percentages is used in combination with chromium to form chrome-molybdenum steel, which has various uses in aircraft. The main alloying ingredient in chrome-molybdenum steel is chrome. Chrome-molybdenum is highly resistant to shock and corrosion.

Nonferrous Aircraft Metals

The term "nonferrous" refers to all metals which have elements other than iron as their base or principal constituent. This group includes such metals asaluminum, titanium, copper, and magnesium, as well as such alloyed metals as Monel and babbit.

Aluminum and Aluminum Alloys

Aluminum is one of the most widely used metals in modern aircraft construction. It is vital to the aviation industry because of its high strength-to-weight ratio and its comparative ease of fabrication. The outstanding characteristic of aluminum is its light weight. Aluminum melts at the comparatively low temperature of 1,250°F. It is nonmagnetic and is an excellent conductor.

The various types of aluminum may be divided into two general classes: (1) The casting alloys (those suitable for casting in sand, permanent mold, or die castings), and (2) the wrought alloys (those which may be shaped by rolling, drawing, or forging). Of these two, the wrought alloys are the most widely used in aircraft construction, being used for stringers, bulkheads, skin, rivets, and extruded sections.

Wrought aluminum and wrought aluminum alloys are divided into two general classes, non heat-treatable alloys and heat-treatable alloys.

Aluminum Alloy Designations

Wrought aluminum and wrought aluminum alloys are designated by a four-digit index system. The system is broken into three distinct groups: 1xxx group, 2xxx through 8xxx group, and 9xxx group (which is at present unused).

In the 2xxx through 8xxx groups, the first digit indicates the major alloying element used in the formation of the alloy as follows:

2xxx—copper

3xxx—manganese

4xxx—silicon

5xxx—magnesium

6xxx—magnesium and silicon

7xxx—zinc

8xxx—other elements

In the 2xxx through 8xxx alloy groups, the second digit in the alloy designation indicates alloy modifications. If the second digit is zero, it indicates the original alloy, while digits 1 through 9 indicate alloy modifications.

The last two of the four digits in the designation identify the different alloys in the group.

Alclad Aluminum

The terms "Alclad and Pureclad" are used to designate sheets that consist of an aluminum alloy core coated with a layer of pure aluminum on each side. The pure aluminum coating affords a dual protection for the core, preventing contact with any corrosive agents, and protecting the core electrolytically by preventing any attack caused by scratching or from other abrasions.

Titanium Alloys

Because titanium is light (it weighs half as much as steel) and strong, particularly when alloyed, it is much in demand for structural parts on high-speed aircraft. Because it remains strong up to 800°F, it is better than aluminum for use around hot sections of jet aircraft.

Titanium resists fatigue, cracking, fracturing, and corrosion. Alloyed, it does not need to be coated to prevent corrosion, but, because it is cathodic to magnesium and aluminum, it should be insulated from them.

Copper Alloys

Copper is one of the most widely distributed metals. It is the only reddish-colored metal and is second only to silver in electrical conductivity. Its use as a structural material is limited because of its great weight. However, some of its outstanding characteristics, such as its high electrical and heat conductivity, in many cases overbalance the weight factor. Because it is very malleable and ductile, copper is ideal for making wire. It is corroded by salt water but is not affected by fresh water. In aircraft, copper is used primarily in the electrical system for bus bars, bonding, and as lockwire.

Nickel Alloys

Monel is a nickel alloy which has the properties of high strength and excellent corrosion resistance. Monel is used for the construction of sprockets and chains for landing gears, in the manufacturing of certain aircraft fasteners, and, in general, whenever both strength and high resistance to corrosion are needed.

Magnesium and Magnesium Alloys

Magnesium, the world's lightest structural metal, is a silvery-white material weighing only two-thirds as much as aluminum. Magnesium does not possess sufficient strength in its pure state for structural uses, but when alloyed with zinc, aluminum, and manganese it produces an alloy having the highest strength-to-weight ratio of any of the commonly used metals.

New Words/ Phrases/ Expression

1. carbon steel 碳钢
2. plaincarbon steel 普通碳钢,碳素钢
3. nickel['nɪkəl] n. & v. [化]镍(Ni);镀镍于
4. elastic limit 弹性极限
5. chromium['krəumiəm] n. [化]铬(Cr)

6. corrosion-resistant　耐蚀

7. antifriction[ˌæntɪˈfrɪkʃən]　*n. & adj.* 减低或防止摩擦之物，润滑剂；减少摩擦的

8. stainless steel　不锈钢

9. vanadium[vəˈneɪdiəm]　*n.* 钒（V）

10. fatigue[fəˈtiːɡ]　*n. & v.* 疲劳，疲乏

11. molybdenum[məˈlɪbdənəm]　*n.* ［化］钼（Mo）

12. aluminum[əˈljuːminəm]　*n.* ［化］铝（Al）

13. titanium[tɪˈteɪniəm]　*n.* ［化］钛（Ti）

14. copper[ˈkɒpə(r)]　*n.* ［化］铜（Cu）

15. magnesium[mæɡˈniːziəm]　*n.* ［化］镁（Mg）

16. Monel[məuˈnel]　*n.* 蒙乃尔铜-镍合金

17. babbit[ˈbæbɪt]　巴比合金，巴氏合金

18. casting alloy　铸造合金

19. wrought alloy　可锻合金，形变合金

20. manganese[ˈmæŋɡəniːz]　*n.* ［化］锰（Mn）

21. silicon[ˈsɪlɪkən]　*n.* ［化］硅，硅元素（Si）

22. zinc[zɪŋk]　*n.* ［化］锌（Zn）

23. alclad aluminum　包铝

24. corrosive agent　腐蚀剂，腐蚀介质

25. electrolytically[elektrəuˈlɪtɪklɪ]　*adv.* 以电解

26. cathodic[kəˈθɒdɪk]　*adj.* 阴极的，负极的

27. insulate[ˈɪnsjuleɪt]　*v.* 使隔离；使绝缘，使隔热

28. silver[ˈsɪlvə(r)]　*n. & adj. & v.* 银；银币；银制品；银色；银制的；银色的；银白色的；
　　　　　　　　　　　　　　（在某物上）镀银；使具有银色光泽，使变成银色

29. electrical conductivity　电导率

30. bus bar　汇流条，母线

31. sprocket[ˈsprɒkɪt]　*n.* 链轮齿

32. landing gear　起落架，起落装置

33. fastener[ˈfɑːsnə(r)]　*n.* 紧固件

Notes

(1) In selecting materials for aircraft repair, many factors are considered in relation to the mechanical and physical properties.

分析：select for 为……而挑选；in relation to 与……有关。

翻译：选择飞机修理的材料时，要考虑到许多与机械和物理性能有关的因素。

(2) Chromium steel is high in hardness, strength, and corrosion-resistant properties, and is particularly adaptable for heat-treated forgings which require greater toughness and strength than may be obtained in plain carbon steel.

分析：be particularly adaptable for 特别适用于；heat-treated forgings 热处理后的锻件；"which require greater toughness and strength than may be obtained in plain carbon steel"作为后置定语修饰 forgings。

翻译：铬钢的硬度、强度和耐腐蚀性能高,特别适用于热处理后的锻件。这类锻件所需要的韧性和强度比碳素钢能提供的要更大。

(3)The pure aluminum coating affords a dual protection for the core, preventing contact with any corrosive agents, and protecting the core electrolytically by preventing any attack caused by scratching or from other abrasions.

分析：contact with 与……有交往(联系)。

翻译：纯铝层为中间层提供双重的保护,防止其接触任何腐蚀剂,并保护核心以防止因为划伤或其他擦伤引起的任何损害而(导致)电解。

(4)Its use as a structural material is limited because of its great weight. However, some of its outstanding characteristics, such as its high electrical and heat conductivity, in many cases overbalance the weight factor.

分析：outstanding characteristics 突出的特点,显著特性；electrical and heat conductivity 电导率和热导率；in many cases 在很多情况下；overbalance 直译为(使)失去平衡而歪倒,过平衡,过重,此处意为压倒,战胜。

翻译：作为一种结构材料它的使用是有限的,因为它的重量很大。然而在许多情况下,它的一些显著特性,譬如高的导电和导热性能,(让我们)忽略了其重量因素。

Exercises

Ⅰ. **Answer the following questions.**

(1)What is the difference between carbon steel and iron?

(2)What are the metals whose principal content is iron called?

(3)What percentages of carbon does low-carbon steel contain?

(4)What is stainless steel produced by alloying?

(5)What is the feature of titanium alloys?

Ⅱ. **Translating the following sentences into Chinese.**

(1)If carbon is added to iron, in percentages ranging up to approximately 1 percent, the product is vastly superior to iron alone and is classified as carbon steel.

(2)Steel containing carbon in percentages ranging from 0.30 to 0.50 percent is classed as medium-carbon steel. This steel is especially adaptable for machining or forging, and where surface hardness is desirable.

(3)The various nickel steels are produced by combining nickel with carbon steel. Nickel increases the hardness, tensile strength, and elastic limit of steel without appreciably decreasing the ductility.

(4)Of these two, the wrought alloys are the most widely used in aircraft construction, being used for stringers, bulkheads, skill, rivets, and extruded sections.

(5) Magnesium, the world's lightest structural metal, is a silvery-white material weighing only two-thirds as much as aluminum.

Ⅲ. Match the first digit with the major alloying element used in the formation of the aluminum alloy by drawing a line.

2xxx	silicon
3xxx	manganese
4xxx	magnesium and silicon
5xxx	zinc
6xxx	copper
7xxx	other elements
8xxx	magnesium

EXTENSIVE READING

Nomenclature and Chemical Compositions of Steels

In order to facilitate the discussion of steels, some familiarity with their nomenclature is desirable. A numerical index, sponsored by the Society of Automotive Engineers (SAE) and the American Iron and Steel Institute (AISI), is used to identify the chemical compositions of the structural steels. In this system, a four-numeral series is used to designate the plain carbon and alloy steels; five numerals are used to designate certain types of alloy steels. The first two digits indicate the type of steel, the second digit also generally (but not always) gives the approximate amount of the major alloying element and the last two (or three) digits are intended to indicate the approximate middle of the carbon range. However, a deviation from the rule of indicating the carbon range is sometimes necessary.

Small quantities of certain elements are present in alloy steels that are not specified as required. These elements are considered as incidental and may be present to the maximum amounts as follows: copper, 0.35 percent; nickel, 0.25 percent; chromium, 0.20 percent; molybdenum, 0.06 percent.

Metal stock is manufactured in several forms and shapes, including sheets, bars, rods, tubings, extrusions, forgings, and castings. Sheet metal is made in a number of sizes and thicknesses. Specifications designate thicknesses in thousandths of an inch. Bars and rods are supplied in a variety of shapes, such as round, square, rectangular, hexagonal, and octagonal. Tubing can be obtained in round, oval, rectangular, or streamlined shapes. The size of tubing is generally specified by outside diameter and wall thickness.

The sheet metal is usually formed cold in such machines as presses, bending brakes, draw benches, or rolls. Forgings are shaped or formed by pressing or hammering heated metal in dies. Castings are produced by pouring molten metal into molds. The casting is finished by machining.

Spark testing is a common means of identifying various ferrous metals. In this test the piece of iron or steel is held against a revolving grinding stone and the metal is identified by the sparks thrown off. Each ferrous metal has its own peculiar spark characteristics. The spark streams vary from a few tiny shafts to a shower of sparks several feet in length. (Few nonferrous metals give off sparks when touched to a grinding stone. Therefore, these metals cannot be successfully identified by the spark test.)

Identification by spark testing is often inexact unless performed by an experienced person.

Effect of Alloying Element in Aluminum Alloys

1000 series. 99% or higher, excellent corrosion resistance, high thermal and electrical conductivity, low mechanical properties, excellent workability. Iron and silicon are major impurities.

2000 series. Copper is the principle alloying element. Solution heat treatment, optimum properties equal to mild steel, poor corrosion resistance unclad. It is usually clad with 6000 or high purity alloy. Its best known alloy is 2024.

3000 series. Manganese is the principle alloying element of this group which is generally non-heat-treatable. The percentage of manganese which will be alloy effective is 1.5%. The most popular is 3003 which is of moderate strength, and has good working characteristics.

4000 series. Silicon is the principle alloying element. This lowers the melting temperature. Its primary use is in welding and brazing. When used in welding heat-treatable alloys, this group will respond to a limited amount of heat treatment.

5000 series. Magnesium is the principle alloying element. It has good welding and corrosion resistant characteristics. High temperatures (over 150℉) or excessive cold working will increase susceptibility to corrosion.

6000 series. Silicon and magnesium form magnesium silicide which makes alloys heat-treatable. It is of medium strength, good forming and has corrosion resistant characteristics.

7000 series. Zinc is the principle alloying element. The most popular alloy of the series is 6061. When coupled with magnesium, it results in heat-treatable alloys of very high strength. It usually has copper and chromium added. The principle alloy of this is 7075.

Copper Alloys

Beryllium copper is one of the most successful of all the copper base alloys. The most valuable feature of this metal is that the physical properties can be greatly stepped up by heat treatment. The resistance of beryllium copper to fatigue and wear makes it suitable for diaphragms, precision bearings and bushings, ball cages, and spring washers.

Brass is a copper alloy containing zinc and small amounts of aluminum, iron, lead, manganese, magnesium, nickel, phosphorous, and tin. Brass with a zinc content of 30 to 35 percent is very ductile, but that containing 45 percent has relatively high strength.

Muntz metal is a brass composed of 60 percent copper and 40 percent zinc. It has excellent corrosion-resistant qualities in salt water. It is used in making bolts and nuts, as well as parts that come in contact with salt water.

Bronzes are copper alloys containing tin. The true bronzes have up to 25 percent tin, but those with less than 11 percent are most useful, especially for such items as tube fittings in aircraft.

Among the copper alloys are the copper aluminum alloys, of which the aluminum bronzes rank very high in aircraft usage. Wrought aluminum bronzes are almost as strong and ductile as medium-carbon steel, and they possess a high degree of resistance to corrosion by air, salt water, and chemicals. They are readily forged, hot- or cold-roiled, and many react to heat treatment. These copper-base alloys contain up to 16 percent of aluminum (usually 5 to 11 percent), to which other metals such as iron, nickel, or manganese may be added. Aluminum bronzes have good tearing qualities, great strength, hardness, and resistance to both shock and fatigue. Because of these properties, they are used for diaphragms, gears, and pumps. Aluminum bronzes are available in rods, bars, plates, sheets, strips, and forgings.

Lesson 4　Heat Treatment

INTENSIVE READING

Heat Treatment of Ferrous Metals

Heat treatment is a series of operations involving the heating and cooling of metals in the solid state. Its purpose is to change a mechanical property or combination of mechanical properties so that the metal will be more useful, serviceable, and safe for a definite purpose. By heat treating, a metal can be made harder, stronger, and more resistant to impact. Heat treating can also make a metal softer and more ductile. No one heat-treating operation can produce all of these characteristics. In fact, some properties are often improved at the expense of others. In being hardened, for example, a metal may become brittle.

The various heat-treating processes are similar in that they all involve the heating and cooling of metals. They differ, however, in the temperatures to which the metal is heated, the rate at which it is cooled, and, of course, in the final result.

The most common forms of heat treatment for ferrous metals are hardening, tempering, normalizing, annealing, and casehardening. Most nonferrous metals can be annealed and many of them can be hardened by heat treatment. However, there is only one nonferrous metal, titanium, that can be casehardened, and none can be tempered or normalized.

The first important consideration in the heat treatment of a steel part is to know its chemical composition. This, in turn, determines its upper critical point. When the upper critical point is known, the next consideration is the rate of heating and cooling to be used. Carrying out these operations involves the use of uniform heating furnaces, proper temperature controls, and suitable quenching mediums.

Hardening

Pure iron, wrought iron, and extremely low-carbon steels cannot be appreciably hardened by heat treatment, since they contain no hardening element. Cast iron can be hardened, but its heat treatment is limited.

In plain carbon steel, the maximum hardness depends almost entirely on the carbon content of the steel. As the carbon content increases, the ability of the steel to be hardened increases. However, this increase in hardenability with an increase in carbon content continues only to a certain point. In practice, that point is 0. 85 percent carbon content. When the carbon content is increased beyond 0. 85 percent, there is no increase in wear resistance.

For most steels, the hardening treatment consists of heating the steel to a temperature just above the upper critical point, soaking or holding for the required length of time, and then cooling it rapidly by plunging the hot steel into oil, water, or brine. Brine is the most severe medium, water is next, and oil is the least severe. Generally an oil quench is used for alloy steels, and brine or water for carbon steels. Hardening increases the hardness and strength of the steel but makes it less ductile.

Tempering

Tempering reduces the brittleness imparted by hardening and produces definite physical properties within the steel. Tempering always follows, never precedes, the hardening operation. In addition to reducing brittleness, tempering softens the steel.

Tempering is always conducted at temperatures below the low critical point of the steel. In this respect, tempering differs from annealing, normalizing, or hardening, all of which require temperatures above the upper critical point.

Generally, the rate of cooling from the tempering temperature has no effect on the resulting structure; therefore, the steel is usually cooled in still air after being removed from the furnace.

Annealing

Annealing of steel produces a fine-grained, soft, ductile metal without internal stresses or strains. In the annealed state, steel has its lowest strength. In general, annealing is the opposite of hardening.

Annealing of steel is accomplished by heating the metal to just above the upper critical point, soaking at that temperature, and cooling very slowly in the furnace. Soaking time is approximately 1 hour per inch of thickness of the material. To produce maximum softness in steel, the metal must be cooled very slowly.

Normalizing

Normalizing of steel removes the internal stresses set up by heat treating, welding, casting, forging, or machining. Stress, if not controlled, will lead to failure. Because of the better physical properties, aircraft steels are often used in the normalized state, but seldom, if ever, in the annealed state.

Normalizing is accomplished by heating the steel above the upper critical point and cooling in still air. The more rapid quenching obtained by air cooling, as compared to furnace cooling, results in a harder and stronger material than that obtained by annealing.

Casehardening

Casehardening produces a hard wear-resistant surface or case over a strong, tough core. Casehardening is ideal for parts which require a wear-resistant surface and, at the same time, must be tough enough internally to withstand the applied loads. The steels best suited to casehardening are the low-carbon and low-alloy steels. If high-carbon steel is casehardened, the hardness penetrates the core and causes brittleness.

In casehardening, the surface of the metal is changed chemically by introducing a high carbide or nitride content. The core is unaffected chemically. When heat treated, the surface responds to hardening while the core toughens. The common forms of casehardening are carburizing, cyaniding, and nitriding. Since cyaniding is not used in aircraft work, only carburizing and nitriding are discussed.

Heat Treatment of Aluminum Alloy

Natural Aging

The heat-treated aluminum alloys are extensively used in aircraft structures. Commonly used heat-treatable alloys naturally age hardened are 2117, 2017 and 2024.

There are three stages in the heat treatment process: heating, quenching, and aging. In its "as fabricated" condition, the aluminum is placed in a heat-treating oven and heated for a period of time determined by its alloying elements and by its size (Figure 4 – 1). Heating time varies according to thickness of metal. Time period between heating and quenching must be accomplished as soon as possible to reduce the possibility of causing intergranular corrosion.

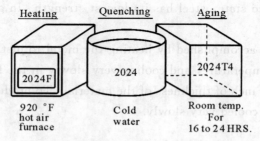

Figure 4 – 1   Heat treatment with natural age hardening

For example, the aluminum alloy 2024 – F is heated to a temperature of 920℉ and held at that temperature until all parts are equally heated. After being heated, the metal is quickly transferred to a quenching tank. The quenching operation must be performed as quickly as possible in order to reduce the possibility of intergranular corrosion, after quenching, the aluminum alloy is kept at room temperature for 16 to 24 hours to ensure that the metal has age hardened. Age hardening, the final step in the heat-treatment process, makes the metal naturally hard.

The designation of metal so treated changes from F, as fabricated, to T4, heat treated and age hardened. If the metal is further hardened by cold working (that is, mechanically) its designation is T3.

Heat-treatable aluminum alloys commonly used in the construction of modern aircraft are 2117 – T4, 2017 – T4 and 2024 – T4 or T3.

The alloy 2117 – T4 is manufactured only as rivet stock. It is the most widely used rivet alloy in the aviation industry. The 2117 – T4 rivet is driven in the condition received from the manufacturer without any further treatment.

Precipitation Heat Treatment: Artificial Aging

Aluminum alloy containing zinc, magnesium, silicon, or copper are given a precipitation heat treatment after natural heat treatment is completed. For example, the alloy 7075 is given a normal heat treatment at 870°F and quenched in cold water. After it is precipitation heat treated at 250°F for 24 hours , it becomes 7075 – T6. Alloys are precipitation heat treated by heating them in an oven; time and temperatures vary.

This treatment has the effect of locking together particles in the grain of the metal, thus increasing strength, stability, and resistance to corrosion. Natural heat treatment begins the grain-binding process; precipitation heat treatment completes it. In addition, artificially aged alloys are generally over-aged to increase their resistance to corrosion, especially if, like 2024, they are subject to intergranular corrosion.

Metals which are given precipitation heat treatment usually lose some malleability and ductility, and their mechanical properties are so changed as to reduce their ability to be reshaped cold without cracking.

The most commonly used precipitation heat-treated alloys are those containing zinc. The alloy 7075 – T6 has high impact resistance and therefore is used where great strength is required. The 7079 – T6 aluminum alloys are excellent for making forgings for heavy channels that carry landing gears or flaps of large aircraft. The alloy 7178 is used where compression loads are the greatest, for example, in the superstructure of wide-body jets.

The alloy 7050 – T73 is the newest aluminum alloy. A combination of aluminum, zinc, and magnesium, it is primarily used as a solid-shank rivet. The 7050 – T73 alloy is the strongest of any rivet alloy in use today. It has a high resistance to stress corrosion and is much stronger than the alloy 2024 – T31, which it has replaced on some modern jetliners.

New Words/ Phrases/ Expression

1. heat treatment　热处理
2. hardening[ˈhɑːdnɪŋ]　v. 淬火；(使)变硬(harden 的现在分词)
3. tempering[ˈtempərɪŋ]　v. 回火；使回火(temper 的现在分词)
4. normalizing[ˈnɔːməlaɪzɪŋ]　n. & v. 常化，正火；(使)正常化(normalize 的现在分词)
5. annealing[əˈniːlɪŋ]　v. 退火(anneal 的现在分词)
6. casehardening[ˈkeɪshɑːdnɪŋ]　v. 表层硬化，使表面硬化(caseharden 的现在分词)
7. critical point　相变点，临界点
8. furnace[ˈfɜːnɪs]　n. 熔炉，火炉
9. quenching medium　淬火介质，淬火剂
10. quench[kwentʃ]　v. 将(热物体)放入水中急速冷却
11. wear resistance　耐磨性
12. soak[səuk]　v. 浸，泡，浸透
13. brine[braɪn]　n. 卤水；浓盐水；海；海水
14. penetrate[ˈpenɪtreɪt]　v. 穿透，渗透
15. carbide[ˈkɑːbaɪd]　n. 碳化物；硬质合金
16. nitride[ˈnaɪtraɪd]　n. 氮化物
17. carburizing[ˈkɑːbjuraɪzɪŋ]　n. 渗碳，增碳剂，渗碳剂
18. cyaniding[ˈsaɪənaɪdɪŋ]　n. 氰化
19. nitriding[ˈnaɪtraɪdɪŋ]　n. 渗氮，表面氮化
20. natural aging　自然时效
21. intergranular corrosion　晶间腐蚀；粒间腐蚀
22. age hardened　时效硬化
23. cold working　冷加工，冷作
24. precipitation heat treatment　沉析热处理
25. artificial aging　人工时效
26. over-aged　过时效，过度失效
27. flap[flæp]　n. 襟翼
28. wide-body jet　宽体客机
29. solid-shank rivet　实心杆铆钉
30. stress corrosion　应力腐蚀

Notes

(1) By heat treating, a metal can be made harder, stronger, and more resistant to impact. Heat treating can also make a metal softer and more ductile. No one heat-treating operation can produce all of these characteristics. In fact, some properties are often

improved at the expense of others. In being hardened，for example，a metal may become brittle.

分析：resistant to impact 耐冲击；at the expense of 在损失（损坏）某事物的情况下。

翻译：通过热处理，金属可以更硬，更强，更耐冲击。热处理也可以使金属更柔软，更有韧性。没有一种热处理操作可以获得所有的这些特性。事实上，（我们）在改善某些金属特性的同时往往损失了其他的特性。例如，金属变硬后可能会变脆。

（2）Casehardening is ideal for parts which require a wear-resistant surface and，at the same time，must be tough enough internally to withstand the applied loads.

分析：be ideal for 是理想的；which 引导了两个后置定语从句"require a wear-resistant surface"和"must be tough enough internally to withstand the applied loads"来修饰先行词 parts；而第二个从句中 be ... enough to do 表示足以……

翻译：有些零件需要耐磨的表面，同时，内部要足够强韧以承受所施加的载荷。对于这类零件，表面硬化处理是理想的。

（3）This treatment has the effect of locking together particles in the grain of the metal，thus increasing strength，stability，and resistance to corrosion. Natural heat treatment begins the grain-binding process；precipitation heat treatment completes it.

翻译：这种处理有锁定晶粒中金属微粒的效果，从而增加强度、稳定性和耐腐蚀性。自然热处理是晶粒开始结合的过程；沉淀热处理使它完成。

（4）The alloy 7178 is used where compression loads are the greatest，for example，in the superstructure of wide-body jets.

翻译：7178 合金使用在压缩载荷最大的地方，例如在宽体客机的上部结构。

（5）The 7050 – T73 alloy is the strongest of any rivet alloy in use today. It has a high resistance to stress corrosion and is much stronger than the alloy 2024 – T31，which it has replaced on some modern jetliners.

分析：of any... 在所有的……当中；resistance to stress corrosion 耐应力腐蚀。

翻译：在今天所使用的所有铆钉合金中 7050 – T73 合金是最强的。它具有很高的抗应力腐蚀性，比 2024 – T31 合金强度大多了，在一些现代的喷气式飞机上已经取代了 2024 – T31 合金。

Exercises

Ⅰ. **Answer the following questions.**

（1）What is the purpose of the heat treatment?

（2）What are the most common forms of heat treatment?

（3）As the carbon content increases, does the ability of the steel to be hardened increase or decrease?

（4）Which type of quenching medium is the least severe when hardening alloy steels?

（5）Is it true that tempering always follows, never precedes, the hardening operation?

（6）What is the difference between tempering and other heat treatment?

（7）What is the purpose of casehardening?

(8) Is high-carbon steel suited to be casehardened? Why?

(9) How many stages are there in the heat treatment process? What are they?

(10) How does the alloy 7075 become 7075 − T6?

II . Translating the following sentences into Chinese.

(1) Pure iron, wrought iron, and extremely low-carbon steels cannot be appreciably hardened by heat treatment, since they contain no hardening element.

(2) In addition to reducing brittleness, tempering softens the steel.

(3) Tempering is always conducted at temperatures below the low critical point of the steel. In this respect, tempering differs from annealing, normalizing, or hardening, all of which require temperatures above the upper critical point.

(4) The steels best suited to casehardening are the low-carbon and low-alloy steels.

(5) The quenching operation must be performed as quickly as possible in order to reduce the possibility of intergranular corrosion, after quenching, the aluminum alloy is kept at room temperature for 16 to 24 hours to ensure that the metal has age hardened.

(6) Metals which are given precipitation heat treatment usually lose some malleability and ductility, and their mechanical properties are so changed as to reduce their ability to be reshaped cold without cracking.

Extensive Reading

In aircraft maintenance and repair, even a slight deviation from design specification, or the substitution of inferior materials, may result in the loss of both lives and equipment. The use of unsuitable materials can readily erase the finest craftsmanship. The selection of the correct material for a specific repair job demands familiarity with the most common physical properties of various metals.

Properties of Metals

Of primary concern in aircraft maintenance are such general properties of metals and their alloys as hardness malleability, ductility, elasticity, toughness, density, brittleness, fusibility, conductivity contraction and expansion, and so forth. These terms are explained to establish a basis for further discussion of structural metals.

Explanation of Terms

Hardness refers to the ability of a metal to resist abrasion, penetration, cutting action, or permanent distortion. Hardness may be increased by cold-working the metal and, in the case of steel and certain aluminum alloys, by heat treatment. Structural parts are often formed from metals in their soft state and are then heat treated to harden them so that the finished shape will be retained. Hardness and strength are closely associated properties of metals.

Brittleness is the property of a metal which allows little bending or deformation without shattering. A brittle metal is apt to break or crack without change of shape. Because structural metals are often subjected to shock loads, brittleness is not a very desirable property. Cast iron, cast aluminum, and very hard steel are examples of brittle metals.

A metal which can be hammered, rolled, or pressed into various shapes without cracking, breaking, or having some other detrimental effect, is said to be malleable. This property is necessary in sheet metal that is worked into curved shapes such as cowlings, fairings, or wingtips. Copper is an example of a malleable metal.

Ductility is the property of a metal which permits it to be permanently drawn, bent, or twisted into various shapes without breaking. This property is essential for metals used in making wire and tubing. Ductile metals are greatly preferred for aircraft use because of their ease of forming and resistance to failure under shock loads. For this reason, aluminum alloys are used for cowl rings, fuselage and wing skin, and formed or extruded parts, such as ribs, spars, and bulkheads. Chrome molybdenum steel is also easily formed into desired shapes. Ductility is similar to malleability.

Elasticity is that property which enables a metal to return to its original shape when the force which causes the change of shape is removed. This property is extremely valuable because it would be highly undesirable to have a part permanently distorted after an applied load was removed. Each metal has a point known as the elastic limit beyond which it cannot be loaded without causing permanent distortion. In aircraft construction, members and parts are so designed that the maximum loads to which they are subjected will not stress them beyond their elastic limits. This desirable property is present in spring steel.

A material which possesses toughness will withstand tearing or shearing and may be stretched or otherwise deformed without breaking. Toughness is a desirable property in aircraft metals.

Density is the weight of a unit volume of a material. In aircraft work, the specified weight of a material per cubic inch is preferred since this figure can be used in determining the weight of a part before actual manufacture. Density is an important consideration when choosing a material to be used in the design of a part in order to maintain the proper weight and balance of the aircraft.

Fusibility is the ability of a metal to become liquid by the application of heat. Metals are fused in welding. Steels fuse around 2,600°F. And aluminum alloys at approximately 1,100°F.

Conductivity is the property which enables a metal to carry heat or electricity. The heat conductivity of a metal is especially important in welding because it governs the amount of heat that will be required for proper fusion. Conductivity of the metal, to a certain extent, determines the type of jig to be used to control expansion and contraction. In aircraft, electrical conductivity must also be considered in conjunction with bonding, to eliminate radio interference.

Contraction and expansion are reactions produced in metals as the result of heating or cooling. Heat applied to a metal will cause it to expand or become larger. Cooling and heating affect the design of welding jigs, castings, and tolerances necessary for hot-rolled material.

Selection Factors

Strength, weight, and reliability are three factors which determine the requirements to be met by any material used in airframe construction and repair. Airframes must be strong and yet as light in weight as possible. There are very definite limits to which increases in strength can be accompanied by increases in weight. An airframe is so heavy that it could not support a few hundred pounds of additional weight would be of little use.

All metals, in addition to having a good strength/weight ratio, must be thoroughly reliable, thus minimizing the possibility of dangerous and unexpected failures. In addition to these general properties, the material selected for a definite application must possess specific qualities suitable for the purpose.

The material must possess the strength required by the dimensions, weight, and use. There are five basic stresses which metals may be required to withstand. These are tension, compression, shear, bending, and torsion.

The relationship between the strength of a material and its weight per cubic inch, expressed as a ratio, is known as the strength/weight ratio. This ratio forms the basis for comparing the desirability of various materials for use in airframe construction and repair. Neither strength nor weight alone can be used as a means of true comparison. In some applications, such as the skin of monocoque structures, thickness is more important than strength, and, in this instance, the material with the lightest weight for a given thickness or gage is best. Thickness or bulk is necessary to prevent buckling or damage caused by careless handling.

Corrosion is the eating away or pitting of the surface or the internal structure of metals. Because of the thin sections and the safety factors used in aircraft design and construction, it would be dangerous to select a material possessing poor corrosion-resistant characteristics.

Another significant factor to consider in maintenance repair is the ability of a material to be formed, bent, or machined to required shapes. The hardening of metals by cold-working or forming is termed work-hardening. If a piece of metal is formed (shaped or bent) while cold, it is said so be cold-worked. Practically all the work an aviation mechanic does on metal is cold-work. While this is convenient, it causes the metal to become harder and more brittle.

If the metal is cold-worked too much, that is, if it is bent back and forth or hammered at the same place too often, it will crack or break. Usually, the more malleable and ductile a metal is, the more cold-working it can stand. Any process which involves controlled heating and cooling of metals to develop certain desirable characteristics (such as hardness, softness, ductility, tensile strength, or refined grain structure) is called heat treatment or heat treating. With steels site term "heat treating" has a broad meaning and includes such processes as annealing, normalizing, hardening, and tempering.

In the heat treatment of aluminum alloys, only two processes are included: (1) The hardening and toughening process, and (2) the softening process. The hardening and toughening process is called heat treating, and the softening process is called annealing.

Aircraft metals are subjected to both shock and fatigue (vibrational) stresses. Fatigue occurs in materials which are exposed to frequent reversals of loading or repeatedly applied loads, if the fatigue limit is reached or exceeded. Repeated vibration or bending will ultimately cause a minute crack to occur at the weakest point. As vibration or bending continues, the crack lengthens until the part completely fails. This is termed shock and fatigue failure. Resistance to this condition is known as shock and fatigue resistance. It is essential that materials used for critical parts be resistant to these stresses.

New Words

1. malleability[ˌmælɪə'bɪlətɪ]　n. 有延展性,有锻塑性,柔韧性
2. ductility[dʌk'tɪlɪtɪ]　n. 展延性,柔软性;韧性;塑性
3. elasticity[ˌiːlæ'stɪsəti]　n. 弹力,弹性
4. toughness[tʌfnəs]　n. 韧性,坚韧,刚性
5. density['densəti]　n. 密度;浓度
6. brittleness['brɪtlnəs]　n. 脆性,脆度
7. fusibility[fjuːzə'bɪlɪtɪ]　n. 熔性,熔度
8. conductivity[ˌkɒndʌk'tɪvɪtɪ]　n. 传导性,传导率,电导率
9. apt[æpt]　adj. 易于……的
10. crack[kræk]　n. 裂缝,裂纹
11. detrimental[ˌdetrɪ'mentl]　adj. 有害的;不利的
12. cowling['kaʊlɪŋ]　n. 飞机引擎罩
13. fairing['feərɪŋ]　n. 整流装置;整流罩
14. wingtip['wɪŋtɪp]　n. (飞机的)翼尖;翼梢
15. tolerance['tɒlərəns]　n. 宽容,容忍;限度;公差;耐受性
16. jig[dʒɪg]　n. 一种夹具

Exercises

I . **Answer the following questions.**

(1) Which property of a metal allows little bending or deformation without shattering?

(2) Which property of a metal enables a metal to return to its original shape when the force which causes the change of shape is removed?

(3) What are the three factors which determine the requirements to be met by any

material used in airframe construction and repair?

(4) What is strength/weight ratio?

(5) Usually, the more malleable and ductile a metal is, the more cold-working it can stand, isn't it?

Ⅱ. Put the following sentences into Chinese.

(1) Of primary concern in aircraft maintenance are such general properties of metals and their alloys as hardness malleability, ductility, elasticity, toughness, density, brittleness, fusibility, conductivity contraction and expansion, and so forth.

(2) Conductivity of the metal, to a certain extent, determines the type of jig to be used to control expansion and contraction. In aircraft, electrical conductivity must also be considered 9 in conjunction with bonding, to eliminate radio interference.

(3) Fatigue occurs in materials which are exposed to frequent reversals of loading or repeatedly applied loads, if the fatigue limit is reached or exceeded.

(4) If the metal is cold-worked too much, that is, if it is bent back and forth or hammered at the same place too often, it will crack or break.

(5) This property is extremely valuable because it would be highly undesirable to have a part permanently distorted after an applied load was removed.

Lesson 5　Composite Materials

INTENSIVE READING

Composite Materials

Introduction

Composites are comprised of two or more different materials for greater strength, where each material maintains its original properties does not dissolve together. A good example of an everyday composite material is concrete. Concrete is made with select amounts of sand, aggregate, and perhaps even glass fiber mixed with cement to bind it together. If the concrete were broken open to view inside, the individual constituents would be visible. The type and quantities of the individual materials can be adjusted to give the resulting concrete better compressive, tensile, and/or flexural properties depending on the application.

We focus on composite laminates made from a combination of fiber reinforcement and a matrix material that binds the fibers together. Most "fiber-reinforced-matrix composites" that we see every day are made with short glass fibers mixed with a polymer or plastic matrix or resin: tubs, showers, sinks, pools, doors, fenders, and various construction materials fall into this category (Figure 5 – 1). Resin is the glue that holds the fibers together and prevents the fibers from buckling when compressed. Resins are usually supplied as liquid formulations in two parts which can be solidified by adding a curing agent. Ideally, a resin about to cure should have a reasonable pot life, good wetting ability, good adhesion and low shrinkage.

Highly loaded composite structures typically use continuous or long-fiber reinforcement that transfers loads along bundles or sheets (or "plies") of fibers arranged to run the length and width of the structure much like the layers in a sheet of plywood. This type of arrangement is typically used in the manufacture of structures such as: boats, bridges, snowboards, bicycle frames, race cars, and aircraft structures, to mention a few.

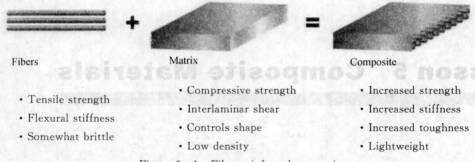

Fibers | Matrix | Composite

- Tensile strength
- Flexural stiffness
- Somewhat brittle

- Compressive strength
- Interlaminar shear
- Controls shape
- Low density

- Increased strength
- Increased stiffness
- Increased toughness
- Lightweight

Figure 5 - 1   Fiber-reinforced composite

Fundamental Composite Material Terminology

Some of the more prominent terms used with composite materials are defined below.

Lamina

A lamina is a flat (or sometimes curved) arrangement of unidirectional (or woven) fibers suspended in a matrix material. A lamina is generally assumed to be orthotropic, and its thickness depends on the material from which it is made. For example, a graphite/epoxy (graphite fibers suspended in an epoxy matrix) lamina may be on the order of 0.005 in (0.127 mm) thick. For the purpose of analysis, a lamina is typically modeled as having one layer of fibers through the thickness. This is only a model and not a true representation of fiber arrangement. Both unidirectional and woven laminas are schematically shown in Figure 5 - 2.

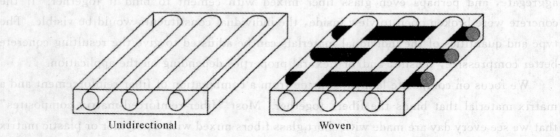

Unidirectional | Woven

Figure 5 - 2   Schematic representation of unidirectional and woven composite lamina

Reinforcements

Reinforcements are used to make the composite structure or component stronger. The most commonly used reinforcements are boron, glass, graphite (often referred to as simply carbon), and Kevlar, but there are other types of reinforcements such as alumina, aluminum, silicon carbide, silicon nitride, and titanium.

Fibers

Fibers are a special case of reinforcements. They are generally continuous and have diameters ranging from 120 to 7,400 μin (3~200 μm). Fibers are typically linear elastic or elastic-perfectly plastic and are generally stronger and stiffer than the same material in bulk form. The most commonly used fibers are boron, glass, carbon, and Kevlar. Fiber and whisker technology is continuously changing.

Matrix

The matrix is the binder material that supports, separates, and protects the fibers. It provides a path by which load is both transferred to the fibers and redistributed among the fibers in the event of fiber breakage. The matrix typically has a lower density, stiffness, and strength than the fibers. Matrices can be brittle, ductile, elastic, or plastic. They can have either linear or nonlinear stress-strain behavior. In addition, the matrix material must be capable of being forced around the reinforcement during some stage in the manufacture of the composite. Fibers must often be chemically treated to ensure proper adhesion to the matrix. The most commonly used matrices are carbon, ceramic, glass, metal, and polymeric. Each has special appeal and usefulness, as well as limitations. The following presentation presents a comprehensive discussion of matrices.

Carbon Matrix

A carbon matrix has a high heat capacity per unit weight. They have been used as rocket nozzles, ablative shields for reentry vehicles, and clutch and brake pads for aircraft.

Ceramic Matrix

A ceramic matrix is usually brittle. Carbon, ceramic, metal, and glass fibers are typically used with ceramic matrices in areas where extreme environments (high temperatures, etc.) are anticipated.

Glass Matrix

Glass and glass-ceramic composites usually have an elastic modulus much lower than that of the reinforcement. Carbon and metal oxide fibers are the most common reinforcements with glass matrix composites. The best characteristic of glass or ceramic matrix composites is their strength at high service temperatures. The primary applications of glass matrix composites are for heat-resistant parts in engines, exhaust systems, and electrical components.

Metal Matrix

A metal matrix is especially good for high-temperature use in oxidizing environments. The most commonly used metals are iron, nickel, tungsten, titanium, magnesium, and aluminum. There are three classes of metal matrix composites:

Class I. The reinforcement and matrix are insoluble (there is little chance that degradation will affect service life of the part). Reinforcement/matrix combinations in this class include tungsten or alumina/copper, BN-coated B or boron/aluminum, and boron/magnesium.

Class II. The reinforcemend/matrix exhibits some solubility (generally over a period of time and during processing) and the interaction will alter the physical properties of the composite. Reinforcemend/matrix combinations included in this class are carbon or tungsted/nickel, tungsted/columbium, and tungsted/copper(chromium).

Class III. The most critical situations in terms of matrix and reinforcement are in this class. The problems encountered here are generally of a manufacturing nature and can be solved through processing controls. Within this class the reinforcement/matrix combinations include alumina or boron or silicon carbide/titanium, carbon or silica/aluminum, and tungsten/copper(titanium).

Polymer Matrix

Polymeric matrices are the most common and least expensive. They are found in nature as amber, pitch, and resin. Some of the earliest composites were layers of fiber, cloth, and pitch. Polymers are easy to process, offer good mechanical properties, generally wet reinforcements well, and provide good adhesion. They are a low-density material. Because of low processing temperatures, many organic reinforcements can be used. A typical polymeric matrix is either viscoelastic or viscoplastic, meaning it is affected by time, temperature, and moisture. The terms thermoset and thermoplastic are often used to identify a special property of many polymeric matrices.

Thermoplastic: a thermoplastic matrix has polymer chains that are not crosslinked. Although the chains can be in contact, they are not linked to each other. A thermoplastic can be remolded to a new shape when it is heated to approximately the same temperature at which it was formed.

Thermoset: a thermoset matrix has highly cross-linked polymer chains. A thermoset can not be remolded after it has been processed. Thermoset matrices are sometimes used at higher temperatures for composite applications.

Laminate

A product made by bonding together two or more layers of materials. In other words, a laminate is a stack of lamina, as illustrated in Figure 5 – 3, oriented in a specific manner to achieve a desired result. Individual lamina is bonded together by a curing procedure that depends on the material system used. The mechanical response of a laminate is different from that of the individual lamina that form it. The laminate's response depends on the properties of each lamina, as well as the order in which the lamina are stacked.

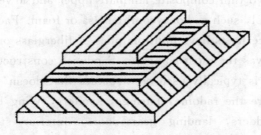

Figure 5 – 3   Schematic of a laminuted composite

Advanced Composites

"Advanced" composites are generally considered to be those that use advanced fiber reinforcements such as carbon fiber and Kevlar, and that exhibit high strength-to-weight ratios, as illustrated in Figure 5 – 4. Advanced composites are strong, stiff, engineered fibers in a high performance resin. They are typically more expensive, with more precisely tailored properties to achieve a specific objective.

Fiberglass vs. Advanced Composites

Some composites are typically referred to as "fiberglass" due to their simple chopped fiberglass and polyester resin composition, whereas most aerospace parts are made using precisely laid plies of carbon fiber/epoxy prepreg.

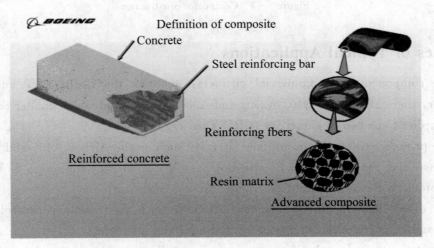

Figure 5 – 4   Schematic of an advanced composite

There are two types of composite construction: Sandwich and Solid laminate, as shown in Figure 5 – 5. In the laminated structures, layers are bonded using a matrix material, multiple layers give thicker cross-section and greater stiffness to the structure, successive layers may have fibers in different directions to bias the load-carrying ability. A sandwich structural formed by two thin composite laminate upper and lower face sheets bonded to a low-density core material, such as honeycomb, balsa or foam. Face plate may be a laminate or sheet metal. The facesheets have multiple plies of fiberglass or carbon. Before the 787, sandwich construction was the most common. Sandwich construction parts have edgebands and sandwich structure is typically used on structure that doesn't carry high loads. Some examples on the 787 are the radome, fairings, spoilers, Wing LE and TE panels, floor panels, elevators, rudders, landing gear doors, interior structure. Solid laminate construction is a composite part constructed without a honeycomb core. Most of the composites on the Boeing 787 are solid laminate construction.

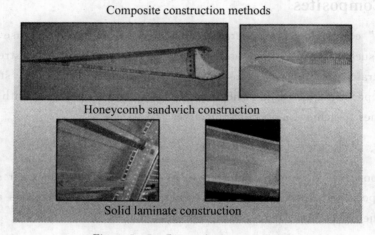

Figure 5 – 5   Composite construction

Examples of Typical Applications

Large components of commercial airliners: such as the Boeing 777 and the 787 Dreamliner, the Airbus A330/340, A350, and A380 aircraft, and many smaller craft such as the Bombardier Canadian Regional Jet (CRJ) (See Figure 5 – 6).

Large primary structures on military aircraft: such as the Airbus A400 and the Boeing C – 17 transports, the B – 2 Spirit Stealth Bomber, V – 22 Osprey tilt-rotor, as well as the F – 22 Raptor and F – 35 JSF fighters.

Many other components on modern airliners: such as radomes, control surfaces, spoilers, landing gear doors, wing-to-body fairings, and interiors.

Large marine vessels and structures: including military and commercial vessels, as well as composite masts (the largest carbon fiber structure in the world is the Mirabella V's 290-foot mast).

Primary components on helicopters: including rotor blades and rotor hubs have been made from carbon fiber and glass epoxy composites since the 1980s. Typically, composites make up 50 to 80 percent of a rotorcraft's airframe by weight. Other parts include radomes, tail cones, and large structural assemblies such as the Fenestron tail rotor on Eurocopter's Dauphin. Bell Helicopter Textron's 429 corporate/EMS/utility helicopter features composite structural sidebody panels, floor panels, bulkheads, nose skins, shroud, doors, fairings, cowlings and stabilizers, most made from carbon fiber/epoxy.

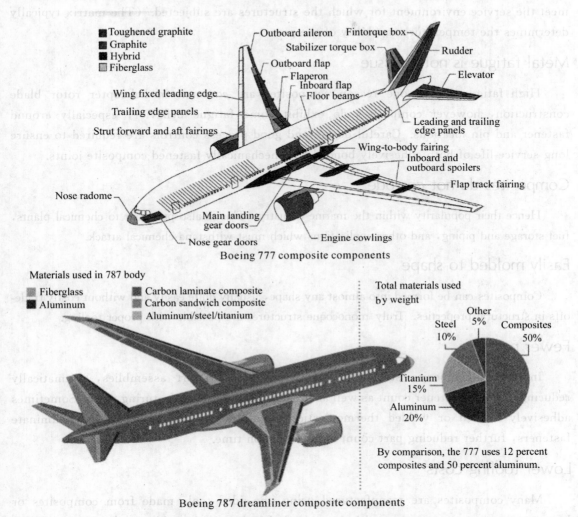

Boeing 777 composite components

Boeing 787 dreamliner composite components

Figure 5 – 6　composite materials used in commercial airliners

Advantages of Composites

High strength and stiffness-to-weight ratio

Composite structures can attain ratios 4 to 10 times better than those made from metals. However, lightweight structures are not automatic. Careful engineering is mandatory, and many tradeoffs are required to achieve truly lightweight structures.

Optimized structures

Fibers are oriented and layers are placed in an engineered stacking sequence to carry specific loads and achieve precise structural performance. Matrix materials are chosen to meet the service environment for which the structures are subjected. (The matrix typically determines the temperature capability of the part.)

Metal fatigue is not an issue

High fatigue life is one reason composites are common in helicopter rotor blade construction, however composites do exhibit some fatigue behavior, especially around fastener and pin locations. Careful design and good process controls are required to ensure long service-life of both adhesively bonded and mechanically fastened composite joints.

Composites do not corrode

Hence their popularity within the marine industry. This is also pertinent to chemical plants, fuel storage and piping, and other applications which must withstand chemical attack.

Easily molded to shape

Composites can be formed into almost any shape, usually quite easily and without costly trade-offs in structural properties. Truly monocoque structures are possible with proper tooling.

Fewer parts

Integrated composite structures often replace multi-part assemblies, dramatically reducing part and fastener count as well as procurement and manufacturing costs. Sometimes adhesively bonded or welded thermoplastic assemblies can almost completely eliminate fasteners, further reducing part count and production time.

Lower tooling costs

Many composites are manufactured using one-sided tools made from composites or Invar, versus more expensive multi-piece metal closed cavity, machine-tooling or large two-sided die sets that are typically required for injection molding of plastics and metal forming processes.

Aerodynamically smooth surfaces

Bonded structures offer smoother surfaces than riveted structures. Composite skins offer increased aerodynamic efficiency, whereas large, thin-skinned metallic structures may exhibit buck-ling between frames under load (i. e. "oil-canning").

"Low observable" or "Stealth" characteristics

Some composite materials can absorb radar and sonar signals and thereby reduce or eliminate "observation" by electronic means. Other materials are "transparent" to radar and work well as a radar "window" in radome applications.

Disadvantages of Composites

Expensive materials

High performance composite materials and processes typically cost more than wood, metal and concrete. The cost of oil and petroleum-based raw material products often drives the price of these materials.

Special storage and handling

Many materials such as film adhesives and prepregs typically have a limited working time (out-time), usually measured in days. Most prepregs also require frozen storage, and may have a limited shelf life of a few months to a year. Material management is critical with these materials.

Not always recyclable

Although there is some ability to recycle thermoplastic composites, recycling of composites in general is not as straight forward as with metals, wood, or unreinforced plastics. Research continues in this area.

Labor-intensive

Tailoring properties typically requires exact material placement, either through hand lay up or automated processes. Each step in the process may require inspection and/or other assessment. This all requires a skilled workforce which can be expensive in some cases.

High capital equipment costs

Ovens, autoclaves, presses, controllers and software are expensive to buy and to operate. Automated machines and programming costs can be a considerable investment.

Easily damaged

Thin-skinned sandwich panels are especially susceptible to damage from low-energy impacts. Such impacts can produce delaminations which are structurally damaging yet invisible to the eye. Moisture intrusion is a special problem in honeycomb sandwich panels, which is also not apparent by visual inspection. Sometimes water intrusion can occur during manufacturing.

Special training and skills are required

A high degree of knowledge and skill is required to properly fabricate and repair advanced composite structures. Good training in this area is mandatory.

Carbon reinforcements can cause galvanic corrosion

Any metal that resides at the anodic end of the corrosion scale will corrode if in direct contact, or fastened with conductive fasteners in close proximity to carbon fiber reinforced structures.

Health and safety concerns

Health issues from working with composite materials may include dermatitis from resins, complications from inhaling respirable fibers, and exposure to suspected carcinogens in some uncured matrix systems. Other safety concerns include transport, storage and disposal of solvents and other materials classified as hazardous.

New Words/Phrases/Expression

1. composite['kəm'pɑːzət] *adj.* & *n.* 混合成的,综合成的合成物,混合物,复合材料
2. laminate['læmə,net] *vt.* & *n.* & *adj.* (将薄片砌合在一起)制成;锻压成薄片层压材料;叠层,层压;由薄片叠成的
3. fiber['faɪbə] *n.* 纤维;光纤;(织物的)质地;纤维物质
4. reinforcement[,riːɪn'fɔːrsmənt] *n.* 加强物;增援;补给品;援军
5. matrix['metrɪks] *n.* 基质;[数]矩阵;模型;母体;子宫
6. polymer['pɑːlɪmə(r)] *n.* 多聚物;[高分子]聚合物
7. resin['rezn] *n.* & *vt.* 树脂;合成树脂;松香用树脂处理;涂擦树脂于
8. lamina['læmənə] *n.* 薄板,薄层,叶片
9. unidirectional[,juːnɪdə'rekʃənl] *adj.* 单向的,单向性的
10. woven['wouvn] *v.* 编,织,织成(weave 的过去分词);编排;杜撰;(把……)编成
11. orthotropic[,ɔːθə'trɒpɪk] *adj.* 正交的
12. graphite['græf,aɪt] *n.* 石墨,黑铅;铅笔粉
13. epoxy[ɪ'pɑːksi] *n.* & *adj.* 环氧树脂;环氧的
14. boron['bɔːrɑːn] *n.* 硼
15. whisker['hwɪskə, 'wɪs-] *n.* 须晶,细须
16. stiffness['stɪfnɪs] *n.* 刚度;僵硬;生硬;强直;顽固
17. ceramic[sə'ræmɪk] *adj.* & *n.* 陶器的,与陶器有关的;陶瓷的,陶制的;陶瓷制品;陶瓷器;制陶艺术;陶瓷装潢艺术
18. tungsten['tʌŋstən] *n.* 钨
19. thermoplastic[,θɜːrmou'plæstɪk] *adj.* & *n.* 热塑性的;热塑性塑料

20. thermoset['θɜːməset]　*n.* & *adj.* 热固树脂,热固塑料;热固的

21. corrode[kə'rəʊd]　*v.* 使腐蚀,侵蚀

22. dermatitis[,dɜːmə'taɪtɪs]　*n.* 皮炎

Notes

(1)Composites are comprised of two or more different materials for greater strength, where each material maintains its original properties does not dissolve together.

分析:where 在此处引导状语从句,修饰句子的主语 Composites。

翻译:复合材料由两种或以上不同的材料组成以获得更高的强度,其中的每种材料保持原来的特性不会融合在一起。

(2)Advanced composites are strong, stiff, engineered fibers in a high performance resin.

翻译:先进的复合材料是高强度、高刚性的工程纤维和高性能树脂组成的复合材料。

(3)Resin is the glue that holds the fibers together and prevents the fibers from buckling when compressed.

翻译:树脂就是胶,它将纤维黏合在一起,防止纤维在压力作用下鼓起、分离。

(4)Resins are usually supplied as liquid formulations in two parts which can be solidified by adding a curing agent. Ideally, a resin about to cure should have a reasonable pot life, good wetting ability, good adhesion and low shrinkage.

翻译:树脂通常为液体状且分两部分提供,加入固化剂后树脂会固化。理想情况下,待固化的树脂应有合理的使用期限以及良好的渗透能力与黏合性,且不易收缩。

(5)Laminate: A product made by bonding together two or more layers of materials.

翻译:层压面板——由两层或两层以上的材料黏合组成。

(6)In the laminated structures, layers are bonded using a matrix material, multiple layers give thicker cross-section and greater stiffness to the structure, successive layers may have fibers in different directions to bias the load-carrying ability.

翻译:在层压结构中,用基质材料把纤维铺层黏合在一起,多层铺层会增加横截面的厚度并增加结构的刚度,连续的几层铺层纤维方向可能不同以满足部件的负载能力。

(7)A sandwich structural formed by two thin composite laminate upper and lower face sheets bonded to a low-density core material, such as honeycomb, balsa or foam. Face plate may be a laminate or sheet metal.

翻译:夹层结构由黏合在低密度芯材(如蜂窝,轻质木材或泡沫塑料)上的上下两薄层复合材料面板构成。面板通常为纤维层压板或金属薄板。

(8)Sandwich structure is typically used on structure that doesn't carry high loads. Some examples on the 787 are the radome, fairings, spoilers, Wing LE and TE panels, floor panels, elevators, rudders, landing gear doors, interior Structure.

翻译:夹层结构主要应用于非承载区域:787上的雷达罩、整流罩、扰流板、机翼前缘和后缘面板、地板面板、升降舵、方向舵、起落架舱门以及内部结构。

Exercises

Ⅰ. **Direction：choose the right meaning for each item.**

(1)增强的纤维

A. glass fibre   B. reinforced fiber   C. synthetic fibre

(2)基质

A. matrix   B. resin   C. laminate

(3)树脂

A. matrix   B. resin   C. laminate

(4)层板

A. matrix   B. resin   C. laminate

(5)金属基复合材料

A. metal matrix composite   B. thermosetting   C. thermoplastic

(6)热固性复合材料

A. metal matrix composite   B. thermosetting   C. thermoplastic

(7)热塑性复合材料

A. metal matrix composite   B. thermosetting   C. thermoplastic

(8)夹层结构

A. sandwich structure   B. solid laminate   C. edgeband

(9)实心层合板结构

A. sandwich structure   B. solid laminate   C. edgeband

(10)织布

A. fabric   B. tape   C. prepreg

(11)单向带

A. fabric   B. tape   C. prepreg

(12)预浸布

A. fabric   B. tape   C. prepreg

Ⅱ. **Fill the right information in the form.**

No.	Term	Abbreviation	Chinese
(1)	Fibrous Composite Material	FCM	
(2)	Carbon Fiber Reinforced Plastic	CFRP	
(3)	Glass Fiber Reinforced Plastic	GFRP	
(4)	Long-Fiber Reinforced Thermoplastic	LFRT	

Ⅲ. **Answer the following questions.**

(1)What's a composite material?

(2)Composite structures differ from metallic structures in several ways. What are they?

(3)What is the Lamina?

(4)What advantages does composite have?

(5)What do sandwich construction and solid laminate construction differ in?

(6)What are the disadvantages of composites?

(7)What is the difference of thermoset and thermoplastic?

(8)How many classes are there of metal matrix composites?

(9)What are the most commonly used matrices?

Ⅳ. Translating the following sentences into Chinese.

(1) Reinforcements are used to make the composite structure or component stronger. The most commonly used reinforcements are boron, glass, graphite (often referred to as simply carbon), and Kevlar, but there are other types of reinforcements such as alumina, aluminum, silicon carbide, silicon nitride, and titanium.

(2)Fibers are a special case of reinforcements. They are generally continuous and have diameters ranging from 120 to 7,400 μin (3~200 μm). Fibers are typically linear elastic or elastic-perfectly plastic and are generally stronger and stiffer than the same material in bulk form. The most commonly used fibers are boron, glass, carbon, and Kevlar. Fiber and whisker technology is continuously changing.

(3) The matrix typically has a lower density, stiffness, and strength than the fibers. Matrices can be brittle, ductile, elastic, or plastic. They can have either linear or nonlinear stress-strain behavior. In addition, the matrix material must be capable of being forced around the reinforcement during some stage in the manufacture of the composite. Fibers must often be chemically treated to ensure proper adhesion to the matrix. The most commonly used matrices are carbon, ceramic, glass, metal, and polymeric. Each has special appeal and usefulness, as well as limitations.

(4) There are two types of composite construction: Sandwich and Solid laminate. Sandwich construction has two thin facesheets around a honeycomb core. The facesheets have multiple plies of fiberglass or carbon. Before the 787, sandwich construction was the most common. Sandwich construction parts have edgebands and sandwich structure is typically used on structure that doesn't carry high loads.

Extensive Reading

Composite Materials in the Airbus A380

Market pressure for a larger aircraft continues to increase. Growth, congestion, economic and environmental factors are driving the need to develop new solutions. Airbus's answer is the A380 family, as shown in Figure 5 - 7. Airbus forecasts a market of about 1,

300 aircraft in the A380 size category over the next 20 years.

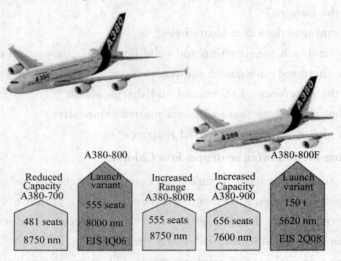

	A380-800		A380-800F
Reduced Capacity A380-700	Launch variant	Increased Range A380-800R	Increased Capacity A380-900	Launch variant
481 seats	555 seats	555 seats	656 seats	150 t
8750 nm	8000 nm	8750 nm	7600 nm	5620 nm
	EIS 1Q06			EIS 2Q08

Figure 5 - 7　The A380 Family

A number of innovations introduced on the A380 will ensure considerable weight savings despite the aircraft's prodigious spaciousness, and countless tests run to date show that aerodynamic performance of the aircraft will also be significantly enhanced. Better aerodynamics and lower airframe weight reduce the demands placed on engines and translate into lower fuel burn, reduced emissions into the atmosphere, and lower operating costs. An estimated 40 percent of the aircraft's structure and components will be manufactured from the latest generation of carbon composites and advanced metallic materials, which, besides being lighter than traditional materials, offer significant advantages in terms of operational reliability, maintainability and ease of repair.

Part of the goal is to select the most appropriate material for the specific application, which would lead to the lightest possible structure. For this purpose, composite materials are good competitors, and their use is foreseen on many areas of the airframe. The weight aspect might be in contradiction to another goal: to standardise material applications across the aircraft or the major assemblies: wing, fuselage … Standardisation plays an important role in manufacture and maintenance over the aircraft's life. So a common understanding of design drivers and maintenance requirements is needed. In parallel, production cost investigations and purchasing activities are also necessary. Thus, material selection is not only driven by structural design criteria.

Inaugural use of a new technology shall proceed step by step, building on experience with earlier Airbus products, as shown in Figure 5 - 8 for composites. The behaviour of structures in service depends not only on material performance, but also on design solutions and manufacturing capabilities. A learning process has to be established for new technologies, which allows for optimisation of materials and processes, increasing areas of application versus time.

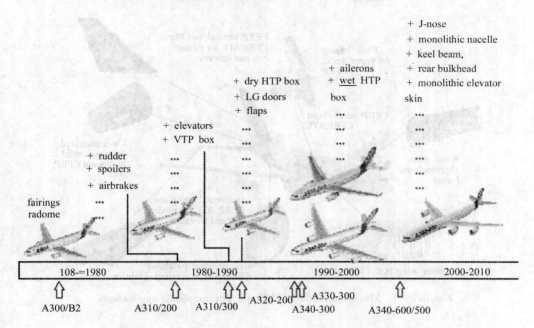

Figure 5 – 8　Evolution of composite material applications at Airbus
HTP—Horizontal Tail Plane；VTP—Vertical Tail Plane；LG—Landing Gear

Due to their mechanical behaviours, design criteria are different for metallic and composite structures. Numerous years of successful experiences at designing metal structure cannot be directly transferred to composite structures. First, composite materials are not isotropic like most metallic alloys. Second, the initiation and growth of damage and the failure modes are more difficult to predict analytically on composites. Due to these complications, the best practices are fully understood only by those engineers that are experienced at designing composite structures. For A380, Airbus benefits from earlier programmes because it was the first manufacturer to make extensive use of composites on large transport commercial aircraft. The A310 was the first production aircraft to have a composite fin box; the A320 was the first aircraft to go into production with an all-composite tail; about 13% by weight of the wing on the A340 is composed of composite materials and the A340/500~600 has CFRP keel beams.

The A380 composite material applications are shown on Figures 5 – 9 and 5 – 10. The A380 will be the first aircraft ever to boast a CFRP (Carbon Fibre Reinforced Plastic) composite central wing box, representing a weight saving of up to one and a half tonnes compared to the most advanced aluminium alloys. On A380 the centre wing box will weigh around 8.8 tonnes, of which 5.3 tonnes is composite materials. The main challenge is the wing root joint, where composite components could be up to 45 mm thick. For this specific application, Airbus will reap a large benefit from the A340 – 600 CFRP keel beams, 16 metres long and 23 mm thick, each of which carries a force of 450 tonnes.

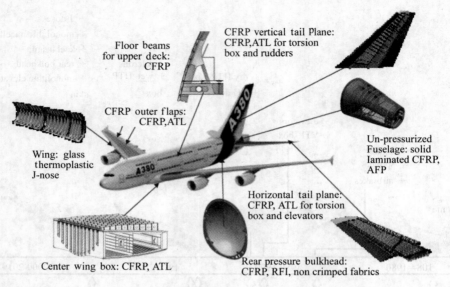

Figure 5 – 9   Major monolithic CFRP and Thermoplastic applications

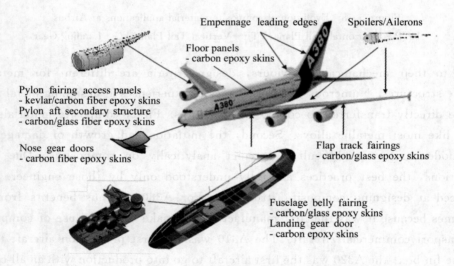

Figure 5 – 10   Major honeycomb applications

A monolithic CFRP design has also been adopted for the fin box and rudder, as well as the horizontal stabilizer and elevators as on previous programmes. Here the main challenge becomes the size of the components. The size of the CFRP Horizontal Tail Plane is close to the size of A320 wings. As for the centre wing box, the size of the components justifies the intensive use of Automated Tape Laying (ATL) technology.

Furthermore, the upper deck floor beams and the rear pressure bulkhead will be made of CFRP. For this last component, different technologies are tested such as Resin Film Infusion (RFI) and Automated Fibre Placement (AFP).

The fixed wing leading edge will be manufactured from thermoplastics, and secondary bracketry in the fuselage (serving, for example, to hold the interior trim) is also likely to be made of thermoplastics. The fixed leading edge (wing-J-nose) in thermoplastics aims at weight and cost savings. This technology has been developed for A340 – 600, demonstrating weight saving, ease of manufacture, improved damage tolerance, and improved inspectability and reparability when compared to the previous A340 metallic D – nose.

Further applications of thermoplastics are under investigation, such as for the ribs in the fixed leading edges of the vertical and horizontal stabilizers.

The choice of CFRP for movable surfaces on the wing trailing edge is regarded to be state-of-the-art. The use of Resin Transfer Moulding (RTM) is foreseen for movable-surface hinges and ribs, when the shape of the components is difficult to obtain using conventional technologies.

Carbon fiber composites

The development of new materials for structural applications has gained attention in recent years due to the rapid growth of the aeronautics industry. In earlier years, metals were the representative material used in airplanes due to their high strength, wide availability, and relatively low cost. However, the heavy weight of metal structural components resulted in high fuel consumption, prompting the search for alternative lighter materials. Constructing a lighter aircraft is therefore a pivotal step towards higher fuel efficiencies, fewer emissions, lower operating costs, and overall sustainability in aerospace. Besides being light weight, aerospace materials must also meet other qualifications. Structural materials in the aerospace industry, especially those involved with the military, must be able to withstand demanding conditions including extremely high temperatures, varying degrees of humidity, and other environmental effects. Additional desirable qualities are long term stability, high impact strength and stiffness to resist damage, and resistance to corrosion and fatigue.

Composites, especially polymer matrix composites (PMCs), became an area of interest after two significant inventions in the 1960s: graphite fibers and boron fibers. Composites made with these fibers have higher longitudinal strength and stiffness compared to the current metallic materials used on aerospace structures. In particular, carbon fibers also have excellent tensile strength, high thermal and electrical conductivities, and good creep resistance. Because of these attractive properties, carbon fibers are in high demand.

In 2006, carbon fiber demand was divided between several industries: 28% aerospace and defense, 50% infrastructure, wind, oil, and gas, and 22% sports goods.

Carbon fibers have been identified as an advantageous reinforcement component in composites. Carbon fiber composites are attractive due to their high stiffness to weight ratio and significant weight savings of 10%～30% compared to typical metallic materials. Carbon fiber composites also have high resistance to environmental and chemical degradation as well as fatigue, resulting in longer service life. Unlike most metals, the qualities of the carbon fiber composite can be chosen and modified to meet the properties required for the application. For example, Li et al. found that oxidizing carbon fibers in nitride matrix composites increased the flexural strength and elastic modulus of the composite. Carbon fiber composites have recently expanded to more primary structural components in aerospace, growing from only 3% usage in 1960 to the 50% used in the revolutionary Boeing 787 airplane. Another unique application of carbon fiber composites is the use in jet engines. The GEnx engine by General Electric, used in the Boeing 787, uses carbon fiber composites for the blades, fan case, and ducts, providing significant weight savings and about 15.4% less fuel use compared to engines used at similar cruising speeds. Besides aerospace, carbon fiber composites are also used in applications such as Formula 1 racecars, sporting goods, and structural materials.

Though advantageous in weight and strength, the use of carbon fiber composites as a structural material has several drawbacks. Even though the strength and stiffness typically exceed that of metals, the low toughness of carbon fiber composites may result in faster and more extreme failures from impacts during the fabrication and operation stages. Carbon fiber composites are also significantly more costly than metals due to material, manufacturing, overhead, and composite processing costs. Environmental factors, such as temperature and humidity, can significantly affect the strength and durability of the polymers. The matrix material, typically a thermoset, is often prone to moisture absorption through a diffusion process which is accelerated at higher humidity. Moisture diffusion is typically a thermally activated process and is modeled by an Arrhenius equation.

Therefore the higher temperatures and longer exposure times that are experienced in many aerospace applications may accelerate moisture absorption. All of these factors must be recognized and considered when using carbon fiber composites as structural components. The selection of the matrix material directly affects the properties of the carbon fiber composite and is therefore critical to the success of the structure. Because carbon fibers do not have significant strength in the transverse direction, a tough matrix material is needed to transfer the load between fibers and provide a strong interfacial bond between the two components. One of the most common matrix materials for carbon fiber composites is epoxy. Epoxy resins typically have high strength and stiffness at a wide range of temperatures, making it a common matrix for PMC's. However, low stability due to relatively low glass transition temperatures prevents the use of epoxies in high temperature applications above 150 ℃. Epoxy resins also have low fracture toughness and are susceptible to delaminations. Due to polarity of the functional groups, epoxy resins can also absorb high amounts of moisture which can further lower the glass transition temperature and undermine the mechanical properties of the resin and resulting composite. For example, on carbon fiber epoxy composites, it was found that the interlaminar shear strength decreased by 25% when 2 wt.% moisture was absorbed. To broaden the conditions under which carbon fiber composites can be used, alternative high temperature matrix materials have been found to replace the epoxy resin.

Lesson 6　Aircraft Structures

INTENSIVE READING

The airframe of a fixed-wing aircraft consists of five principal units: the fuselage, wings, stabilizers, flight control surfaces, and landing gear.

Airframe structural components are constructed from a wide variety of materials. The earliest aircraft were constructed primarily of wood. Steel tubing and the most common material, aluminum, followed. Many newly certified aircraft are built from molded composite materials, such as carbon fiber. Structural members of an aircraft's fuselage include stringers, longerons, ribs, bulkheads, and more. The main structural member in a wing is called the wing spar.

The skin of aircraft can also be made from a variety of materials, ranging from impregnated fabric to plywood, aluminum, or composites. Under the skin and attached to the structural fuselage are the many components that support airframe function. The entire airframe and its components are joined by rivets, bolts, screws, and other fasteners. Welding, adhesives, and special bonding techniques are also used.

In designing an aircraft, every square inch of wing and fuselage, every rib, spar, and even each metal fitting must be considered in relation to the physical characteristics of the metal of which it is made. Every part of the aircraft must be planned to carry the load to be imposed upon it. The determination of such loads is called stress analysis. Although planning the design is not the function of the aviation mechanic, it is, nevertheless, important that he understand and appreciate the stresses involved in order to avoid changes in the original design through improper repairs.

There are five major stresses to which all aircraft are subjected: tension compression torsion, shear, bending (Figure 6 – 1).

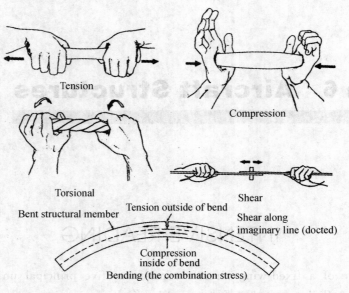

Figure 6 - 1 Five stresses acting on an aircraft

Fixed-Wing Aircraft Fuselage

The fuselage is the main structure or body of the fixed-wing aircraft. It provides space for cargo, controls, accessories, passengers, and other equipment. In single-engine aircraft, the fuselage houses the powerplant. In multiengine aircraft, the engines may be either in the fuselage, attached to the fuselage, or suspended from the wing structure. There are three general types of fuselage construction: truss, monocoque and semi-monocoque.

Truss Type

The truss type fuselage frame (Figure 6 - 2) is usually constructed of steel tubing welded together in such a manner that all members of the truss can carry both tension and compression loads. In some aircraft, principally the light, single-engine models, truss fuselage frames are constructed of aluminum alloy and may be riveted or bolted into one piece, with cross-bracing achieved by using solid rods or tubes.

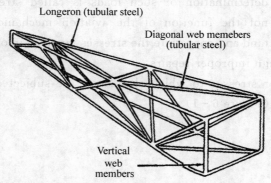

Figure 6 - 2 Warren truss of welded turbular steel

Monocoque Type

The true monocoque (Figure 6 – 3) construction uses formers, frame assemblies, and bulkheads to give shape to the fuselage. The heaviest of these structural members are located at intervals to carry concentrated loads and at points where fittings are used to attach other units such as wings, powerplants, and stabilizers. Since no other bracing members are present, the skin must carry the primary stresses and keep the fuselage rigid. Thus, the biggest problem involved in monocoque construction is maintaining enough strength while keeping the weight within allowable limits.

To overcome the strength/weight problem of monocoque construction, a modification called semi-monocoque construction was developed (Figure 6 – 4).

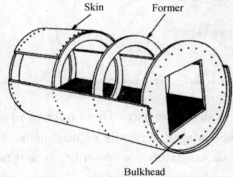

Figure 6 – 3   Monocoque construction   Figure 6 – 4   Semi-monocoque construction

In addition to formers, frame assemblies, and bulkheads, the semi-monocoque construction has the skin reinforced by longitudinal members. The reinforced shell has the skin reinforced by a complete framework of structural members. Different portions of the same fuselage may belong to any one of the three classes, but most aircraft are considered to be of semi-monocoque type construction.

Semi-monocoque Type

The semi-monocoque fuselage is constructed primarily of the alloys of aluminum and magnesium, although steel and titanium are found in areas of high temperatures. Primary bending loads are taken by the longerons, which usually extend across several points of support. The longerons are supplemented by other longitudinal members, called stringers. It also consists of frame assemblies, bulkheads, and formers as used in the monocoque design but, additionally, the skin is reinforced by longitudinal members called longerons. Longerons usually extend across several frame members and help the skin support primary bending loads. They are typically made of aluminum alloy either of a single piece or a built-up construction.

Stringers are also used in the semi-monocoque fuselage. These longitudinal members are typically more numerous and lighter in weight than the longerons. They come in a variety of shapes and are usually made from single piece aluminum alloy extrusions or formed aluminum. Stringers have some rigidity but are chiefly used for giving shape and for attachment of the skin. Stringers and longerons together prevent tension and compression from bending the fuselage (Figure 6 – 5).

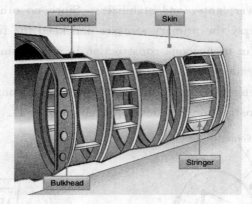

Figure 6 – 5   The most common airframe construction is semi-monocoque

Other bracing between the longerons and stringers can also be used. Often referred to as web members, these additional support pieces may be installed vertically or diagonally. It must be noted that manufacturers use different nomenclature to describe structural members. For example, there is often little difference between some rings, frames, and formers. One manufacturer may call the same type of brace a ring or a frame. Manufacturer instructions and specifications for a specific aircraft are the best guides.

The semi-monocoque fuselage is constructed primarily of alloys of aluminum and magnesium, although steel and titanium are sometimes found in areas of high temperatures. Individually, no one of the aforementioned components is strong enough to carry the loads imposed during flight and landing. But, when combined, those components form a strong, rigid framework. This is accomplished with gussets, rivets, nuts and bolts, screws, and even friction stir welding. A gusset is a type of connection bracket that adds strength.

To summarize, in semi-monocoque fuselages, the strong, heavy longerons hold the bulkheads and formers, and these, in turn, hold the stringers, braces, web members, etc. All are designed to be attached together and to the skin to achieve the full strength benefits of semi-monocoque design. It is important to recognize that the metal skin or covering carries part of the load. The fuselage skin thickness can vary with the load carried and the stresses sustained at a particular location.

The advantages of the semi-monocoque fuselage are many. The bulkheads, intercostal, keel beam, frames, stringers, and longerons facilitate the design and construction of a streamlined fuselage that is both rigid and strong. Spreading loads among these structures and the skin means no single piece is failure critical. This means that a semi-monocoque fuselage, because of its stressed-skin construction, may withstand considerable damage and still be strong enough to hold together.

Fuselages are generally constructed in two or more sections. On small aircraft, they are generally made in two or three sections, while larger aircraft may be made up of as many as six sections or more before being assembled.

Most modern aircraft are considered to be of semi-monocoque type construction. The fuselage is a semi-monocoque structure. Most of material in the fuselage is aluminum. These auxiliary structures attach to the fuselage: Nose radome, Wing-to-body fairing and Tailcone.

These dimensions give locations on the fuselage. The scale for each dimension is inches. The body station line (STA) is a horizontal dimension. It starts at station line zero. You measure the body station line from a vertical reference plane that is forward of the airplane. The body buttock line (BL) is a lateral dimension. You measure the buttock line to the left or right of the airplane center line. The water line (WL) is a height dimension. You measure the water line from a horizontal reference plane below the airplane.

The nose radome is an aerodynamic fairing on the front of the fuselage. Most of the material in the radome is fiberglass. The radome area has navigation and weather radar antennas. Lightning diverter strips prevent damage to these antennas and the equipment they connect to. The lightning diverter strips decrease lightning energy and transmit it to the airframe.

Wings Structure

The wings of an aircraft are designed to lift it into the air. Their particular design for any given aircraft depends on a number of factors, such as size, weight, use of the aircraft, desired speed in flight and at landing, and desired rate of climb. The wings of aircraft are designated left and right, corresponding to the left and right sides of the operator when seated in the cockpit.

Most of material in the wing is aluminum. These components attach to the wing structure: Engine nacelle/pylon, Flight control surfaces, Wingtips, Winglets. The wing has two reference dimensions. The reference dimensions give wing locations in inches. Measure each location from buttock line 0. These are the wing reference dimensions: Wing station and Wing buttock line. Measure the wing station perpendicular to the wing leading edge. Measure the wing buttock line parallel to the buttock line.

Modern aircraft are tending toward lighter and stronger materials throughout the airframe and in wing construction. Wings made entirely of carbon fiber or other composite materials exist, as well as wings made of a combination of materials for maximum strength to weight performance.

The internal structures of most wings are made up of spars and stringers running spanwise and ribs and formers or bulkheads running chordwise (leading edge to trailing edge). The spars are the principle structural members of a wing. They support all distributed loads, as well as concentrated weights such as the fuselage, landing gear, and engines. The skin, which is attached to the wing structure, carries part of the loads imposed during flight. It also transfers the stresses to the wing ribs. The ribs, in turn, transfer the loads to the wing spars(Figure 6 - 6).

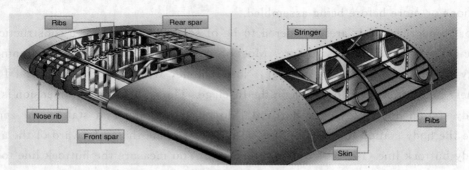

Figure 6 - 6   Wing structure nomenclature

In general, wing construction is based on one of three fundamental designs: (1) Monospar; (2)Multispar;(3)Box beam.

Modification of these basic designs may be adopted by various manufacturers. The monospar wing incorporates only one main spanwise or longitudinal member in its construction. Ribs or bulkheads supply the necessary contour or shape to the airfoil. Although the strict monospar wing is not common, this type of design modified by the addition of false spars or light shear webs along the trailing edge for support of control surfaces is sometimes used.

The multispar wing incorporates more than one main longitudinal member in its construction. To give the wing contour, ribs or bulkheads are often included.

The box beam type of wing construction uses two main longitudinal members with connecting bulkheads to furnish additional strength and to give contour to the wing(Figure 6 - 7). A corrugated sheet may be placed between the bulkheads and the smooth outer skin so that the wing can better carry tension and compression loads. In some cases, heavy longitudinal stiffeners are substituted for the corrugated sheets. A combination of corrugated sheets on the upper surface of the wing and stiffeners on the lower surface is sometimes used. Air transport category aircraft often utilize box beam wing construction.

Wing Spars

Spars are the principal structural members of the wing. They correspond to the longerons of the fuselage. They run parallel to the lateral axis of the aircraft, from the fuselage toward the tip of the wing, and are usually attached to the fuselage by wing fittings, plain beams, or a truss.

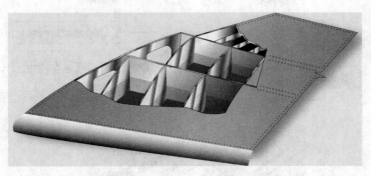

Figure 6 – 7   Box beam construction

Spars may be made of metal, wood, or composite materials depending on the design criteria of a specific aircraft. Currently, most manufactured aircraft have wing spars made of solid extruded aluminum or aluminum extrusions riveted together to form the spar. The increased use of composites and the combining of materials should make airmen vigilant for wings spars made from a variety of materials. Figure 6 – 8 shows examples of metal wing spar cross-sections.

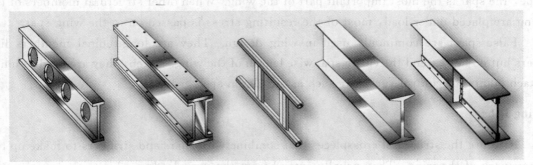

Figure 6 – 8   Examples of metal wing spar shapes

In an I-beam spar, the top and bottom of the I-beam are called the caps and the vertical section is called the web. The entire spar can be extruded from one piece of metal but often it is built up from multiple extrusions or formed angles. The web forms the principal depth portion of the spar and the cap strips (extrusions, formed angles, or milled sections) are attached to it. Together, these members carry the loads caused by wing bending, with the caps providing a foundation for attaching the skin.

Additionally, fail-safe spar web design exists. Fail-safe means that should one member of a complex structure fail, some other part of the structure assumes the load of the failed member and permits continued operation. A spar with failsafe construction is shown in Figure 6 – 9. This spar is made in two sections. The top section consists of a cap riveted to the upper web plate. The lower section is a single extrusion consisting of the lower cap and web plate. These two sections are spliced together to form the spar. If either section of this type of spar breaks, the other section can still carry the load. This is the fail-safe feature.

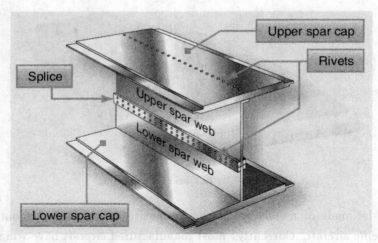

Figure 6 - 9　A fail-safe spar with a riveted spar web

As a rule, a wing has two spars. One spar is usually located near the front of the wing, and the other about two-thirds of the distance toward the wing's trailing edge. Regardless of type, the spar is the most important part of the wing. When other structural members of the wing are placed under load, most of the resulting stress is passed on to the wing spar.

False spars are commonly used in wing design. They are longitudinal members like spars but do not extend the entire spanwise length of the wing. Often, they are used as hinge attach points for control surfaces, such as an aileron spar.

Wing Ribs

Ribs are the structural crosspieces that combine with spars and stringers to make up the framework of the wing. They usually extend from the wing leading edge to the rear spar or to the trailing edge of the wing. The ribs give the wing its cambered shape and transmit the load from the skin and stringers to the spars. Similar ribs are also used in ailerons, elevators, rudders, and stabilizers.

Wing ribs are usually manufactured from either wood or metal. Aircraft with wood wing spars may have wood or metal ribs while most aircraft with metal spars have metal ribs.

A wing rib may also be referred to as a plain rib or a main rib. Wing ribs with specialized locations or functions are given names that reflect their uniqueness. For example, ribs that are located entirely forward of the front spar that are used to shape and strengthen the wing leading edge are called nose ribs or false ribs. False ribs are ribs that do not span the entire wing chord, which is the distance from the leading edge to the trailing edge of the wing. Wing butt ribs may be found at the inboard edge of the wing where the wing attaches to the fuselage. Depending on its location and method of attachment, a butt rib may also be called a bulkhead rib or a compression rib if it is designed to receive compression loads that tend to force the wing spars together.

At the inboard end of the wing spars provide a strong and secure method for attaching the wing to the fuselage. The interface between the wing and fuselage is often covered with a fairing to achieve smooth airflow in this area. The fairing(s) can be removed for access to the wing attach fittings.

The wing tip is often a removable unit, bolted to the outboard end of the wing panel. One reason for this is the vulnerability of the wing tips to damage, especially during ground handling and taxiing. The wing tip assembly is of aluminum alloy construction. The wing tip cap is secured to the tip with countersunk screws and is secured to the interspar structure at four points with-inch diameter bolts. To prevent ice from forming on the leading edge of the wings of large aircraft, hot air from an engine is often channeled through the leading edge from wing root to wing tip. A louver on the top surface of the wingtip allows this warm air to be exhausted overboard. Wing position lights are located at the center of the tip and are not directly visible from the cockpit. As an indication that the wing tip light is operating, some wing tips are equipped with a Lucite rod to transmit the light to the leading edge.

Wing Skin

Often, the skin on a wing is designed to carry part of the flight and ground loads in combination with the spars and ribs. This is known as a stressed-skin design. The all-metal, full cantilever wing section illustrated in Figure 6 – 10 shows the structure of one such design. The lack of extra internal or external bracing requires that the skin share some of the load. Notice the skin is stiffened to aid with this function.

Figure 6 – 10   The skin is an integral load carrying part of a stressed skin design

Fuel is often carried inside the wings of a stressed-skin aircraft. The joints in the wing can be sealed with a special fuel resistant sealant enabling fuel to be stored directly inside the structure. This is known as wet wing design. Alternately, a fuel-carrying bladder or tank can be fitted inside a wing. Figure 6 – 11 shows a wing section with a box beam structural design such as one that might be found in a transport category aircraft. This structure increases strength while reducing weight. Proper sealing of the structure allows fuel to be stored in the box sections of the wing.

The wing skin on an aircraft may be made from a wide variety of materials such as fabric, wood, or aluminum. But a single thin sheet of material is not always employed.

Chemically milled aluminum skin can provide skin of varied thicknesses.

On aircraft with stressed-skin wing design, honeycomb structured wing panels are often used as skin. A honeycomb structure is built up from a core material resembling a bee hive's honeycomb which is laminated or sandwiched between thin outer skin sheets. They have a variety of uses on the aircraft, such as floor panels, bulkheads, and control surfaces, as well as wing skin panels.

Sealed structure fuel tank—wet wing

Figure 6 – 11   Fuel is often carried in the wings

A honeycomb panel can be made from a wide variety of materials. Aluminum core honeycomb with an outer skin of aluminum is common. But honeycomb in which the core is an Arimid fiber and the outer sheets are coated Phenolic is common as well. In fact, a myriad of other material combinations such as those using fiberglass, plastic, Nomex, Kevlar, and carbon fiber all exist. Each honeycomb structure possesses unique characteristics depending upon the materials, dimensions, and manufacturing techniques employed.

Nacelles/pylons

The engines are attached to the wing with struts (pylons). The strut is a pathway for all the pneumatic, electric, fuel, and hydraulic connections to the engine. The strut transfers loads from the engine and thrust reverser to the wing.

Struts are essentially frame and skin structures attached by fasteners to form a torque box. Strut skins are aluminum. Bulkheads, mid and lower spars are made of steel. Stainless steel or titanium is used where high strength, heat and fire resistance are required.

Struts are attached to the wing by means of the upper link between the strut upper spar and wing, the diagonal brace between the strut lower spar and wing, and the strut fittings connecting strut and wing directly. Side braces between strut midspar fittings and wing adds stability. Structural fuse pins or fuse bolts are used at the upper link, diagonal brace and midspar attach points.

Struts transmit engine loads to the wing. The engine is attached at the strut at the forward and aft engine mounts and the thrust link fitting. The forward and aft engine mount transmit vertical and thrust loads to the strut. Torsional loads are transmitted through thrust links connected between the engine fan frame and strut lower spar.

The engine and strut have outer skins which provide smooth airflow. The engine has a nacelle which is a streamlined enclosure. The nacelle consists of the inlet cowl, fan cowl, and thrust reverser. The strut has fairings which provide a smooth, contoured surface between the engine nacelle and wing.

Nacelles are streamlined enclosures used primarily to house the engine and its components. They usually present a round or elliptical profile to the wind thus reducing aerodynamic drag. On most single-engine aircraft, the engine and nacelle are at the forward end of the fuselage. On multiengine aircraft, engine nacelles are built into the wings or attached to the fuselage at the empennage (tail section).

The framework of a nacelle usually consists of structural members similar to those of the fuselage. Lengthwise members, such as longerons and stringers, combine with horizontal/vertical members, such as rings, formers, and bulkheads, to give the nacelle its shape and structural integrity. A firewall is incorporated to isolate the engine compartment from the rest of the aircraft. This is basically a stainless steel or titanium bulkhead that contains a fire in the confines of the nacelle rather than letting it spread throughout the airframe.

Engine mounts are also found in the nacelle. These are the structural assemblies to which the engine is fastened. They are usually constructed from chrome/molybdenum steel tubing in light aircraft and forged chrome/nickel/molybdenum assemblies in larger aircraft.

The exterior of a nacelle is covered with a skin or fitted with a cowling which can be opened to access the engine and components inside. Both are usually made of sheet aluminum or magnesium alloy with stainless steel or titanium alloys being used in high-temperature areas, such as around the exhaust exit. Regardless of the material used, the skin is typically attached to the framework with rivets.

Cowling refers to the detachable panels covering those areas into which access must be gained regularly, such as the engine and its accessories. It is designed to provide a smooth airflow over the nacelle and to protect the engine from damage. Cowl panels are generally made of aluminum alloy construction. However, stainless steel is often used as the inner skin aft of the power section and for cowl flaps and near cowl flap openings. It is also used for oil cooler ducts. Cowl flaps are moveable parts of the nacelle cowling that open and close to regulate engine temperature.

A turbojet engine nacelle, the cowl panels are a combination of fixed and easily removable panels which can be opened and closed during maintenance. A nose cowl is also a feature on a jet engine nacelle. It guides air into the engine.

Empennage

The empennage of an aircraft is also known as the tail section. Most empennage designs consist of a tail cone, fixed aerodynamic surfaces or stabilizers, and movable aerodynamic surfaces.

The tail cone serves to close and streamline the aft end of most fuselages. The cone is made up of structural members like those of the fuselage; however, cones are usually of lighter construction since they receive less stress than the fuselage.

The other components of the typical empennage are of heavier construction than the tail cone. These members include fixed surfaces that help stabilize the aircraft and movable surfaces that help to direct an aircraft during flight. The fixed surfaces are the horizontal stabilizer and vertical stabilizer. The movable surfaces are usually a rudder located at the aft edge of the vertical stabilizer and an elevator located at the aft edge the horizontal stabilizer (Figure 6 – 12).

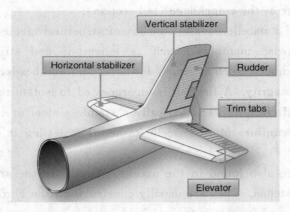

Figure 6 – 12 Components of a typical empennage

The structure of the stabilizers is very similar to that which is used in wing construction. Figure 6 – 13 shows a typical vertical stabilizer. Notice the use of spars, ribs, stringers, and skin like those found in a wing. They perform the same functions shaping and supporting the stabilizer and transferring stresses. Bending, torsion, and shear created by air loads in flight pass from one structural member to another. Each member absorbs some of the stress and passes the remainder on to the others. Ultimately, the spar transmits any overloads to the fuselage. A horizontal stabilizer is built the same way.

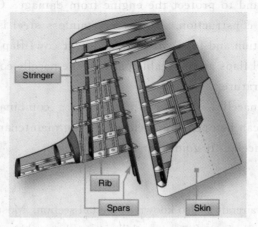

Figure 6 – 13 Vertical stabilizer

The rudder and elevator are flight control surfaces that are also part of the empennage. Flight control surfaces are usually similar in construction to one another and vary only in size, shape, and methods of attachment. On aluminum light aircraft, their structure is often similar to an all-metal wing. This is appropriate because the primary control surfaces are simply smaller aerodynamic devices. They are typically made from an aluminum alloy structure built around a single spar member or torque tube to which ribs are fitted and a skin is attached. Flight control surfaces constructed from composite materials are also commonly used. These are found on many heavy and high-performance aircraft, as well as gliders, and light-sport aircraft. The weight and strength advantages over traditional construction can be significant. A wide variety of materials and construction techniques are employed.

Windows

The windows on the airplane are grouped as follows:
(1) Flight Compartment Windows;
(2) Passenger Cabin Windows;
(3) Door-Mounted Windows;
(4) Viewers and Observation Windows.

All windows are designed to withstand cabin pressurization loads, and are designed with fail-safe features.

Flight Compartment Windows

There are six (ten) windows symmetrically located around the flight compartment. They are named and numbered as shown in Figure 6 - 14. Windows No. 1, 3, 4 and 5 are fixed in place. Window No. 2 is a sliding window, mounted on tracks, to permit ventilation and communication on the ground. The window seals which are used on the flight compartment windows consist of fixed window pressure seals, which are used on windows No. 1, 3, 4 and 5, and the sliding window pressure seals on windows No. 2. The seals prevent cabin pressurization leakage around the window panes when the flight compartment is pressurized. The sealants that are used on the windows prevent moisture penetration, water entrapment, and provide aerodynamic flushness of the outer window pane and the window frame.

Passenger Cabin Windows

Passenger cabin windows are located between the fuselage frames in those areas of the cabin where passenger seating is provided. The passenger cabin windows consist of outer and middle structural panes. The innermost pane is nonstructural and is mounted in the cabin sidewall lining. The outer and middle structural panes are each capable of taking the full cabin pressurization load. Fail-safe structure is ensured by the middle pane which is designed for 1.5 times the normal operating pressure at 70 degrees Fahrenheit. The outer pane is made of stretched acrylic for increased strength. The middle pane is made of a cast acrylic. The inner non-structural pane in the sidewall lining is made from polycarbonate.

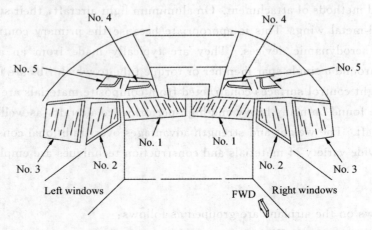

Figure 6 – 14　Windows are located around the flight compartment

The windows are designed as plug type windows subject only to the pressure acting on them. The outer pane of stretched acrylic plastic is rectangular in shape with rounded corners and a beveled outer edge to fit the window frame. The pane is curved to fair with the fuselage contour. The middle pane made of modified acrylic plastic sheet is similarly shaped like the outer pane, but with an unbeveled edge. The middle pane is contained in the window seal. A small breather hole is located near the bottom of the middle pane. The window seal is a molded ethylene propylene seal with six alignment tabs, staggered beads, and an integral masking feature which is removed following installation. Bonding is not required on installation of the seal.

Windows are designed to preclude fogging and frosting by means of multiple pane construction with intervening cavities essentially isolated from cabin interior air conditions. A passenger cabin window plate (plug) may be installed in place of the outer and middle panes in those areas of the passenger cabin where seating is not provided.

Door-Mounted Windows

The windows in the doors consist of two panes and a dust cover. The outer and inner panes are structural. The dust cover is nonstructural and is mounted in the cabin sidewall lining. The outer and inner structural panes are each capable of taking the full cabin pressurization load. Fail-safe structure is ensured by the inner pane which is designed for 1. 5 times the normal operating pressure at 70 degrees Fahrenheit. The outer pane is made of stretched acrylic for increased strength. The inner pane is made of a cast acrylic. The innermost pane in the sidewall lining is made from polycarbonate.

Viewers and Observation Windows

The inspection and observation equipment consists of main gear downlock viewer and a nose gear downlock viewer. The main gear downlock viewer is located in the aisleway of the main cabin floor over the wheel well area and is arranged to provide in-flight visual inspection of the main gear downlock indicators when the main gear is down and locked. The nose gear downlock viewer is installed above the nose gear wheel well and is arranged to provide in-flight visual inspection of the nose gear drag link locking components when the nose gear is down and locked.

Doors

The doors are removable units which provide access to parts of the airplane. The doors are divided into these groups(Figure 6 – 15):

(1) Passenger/Crew Doors;

(2) Emergency Exits;

(3) Cargo Doors;

(4) Access Doors;

(5) Service Doors;

(6) Fixed Interior Doors;

(7) Airplanes with Forward Airstairs; Entrance Stairs.

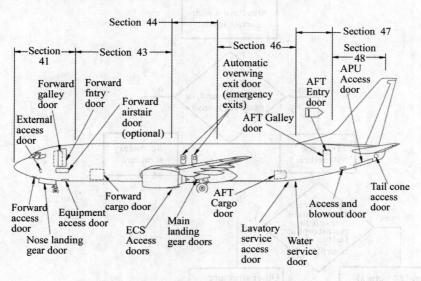

Figure 6 – 15　The door location diagram

The door location diagram gives you the reference system to find the main entry doors (LH and RH side), automatic over-wing exit door (emergency exit), forward and aft cargo doors, service and access doors (forward access door, equipment access door, external/power receptacle door, toilet and water service doors, ground air conditioning access door, access and blowout doors, auxiliary power unit access door, tailcone access door) and the landing gear doors.

Structural Classification

Primary Structure

Primary structure means that which carries flight, ground, or pressure loads. Primary structure is classified into two categories: Principal Structural Elements (PSE) and Other Structure. Most of the primary structures on the airplane are Principal Structural Elements (PSE). PSEs are also known as Structural Significant Items (SSI).

(a) Principal Structural Elements (PSE): Primary structure which contribute significantly to carrying flight, ground, and pressurization loads, and whose failure could result in the catastrophic failure of the airplane. The failure of PSE'S could result in the catastrophic failure of the airplane.

(b) Other Structure: Primary structure that is not a Principal Structural Element (PSE).

Secondary Structure

Structure which carries only air or inertial loads generated on or within the secondary structure. Most secondary structures are important to the aerodynamic performance of the airplane.

Figure 6 – 16 shows how to judge the aircraft structure classification.

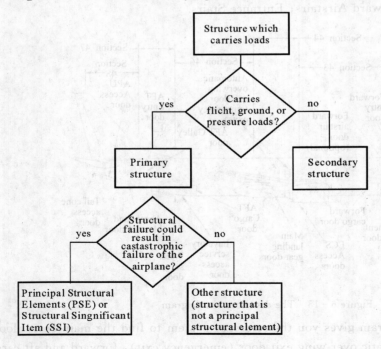

Figure 6 – 16   Step of structure classification

New Words/Phrases/Expression

1. principal['prɪnsəp(ə)l]　*adj.* & *n.* 主要的;资本的;首长;校长;资本;当事人
2. stabilizer['steɪbɪlaɪzə]　*n.* 稳定器;安定装置
3. component[kəm'pəʊnənt]　*adj.* & *n.* 组成的,构成的;成分;组件;[电子]元件
4. carbon fiber　碳化纤维,[材]碳素纤维
5. longeron['lɒn(d)ʒərən]　*n.* (飞机的)纵梁
6. impregnate['ɪmpregneɪt]　*vt.* & *adj.* 灌输;浸透;充满的
7. plywood['plaɪwʊd]　*n.* 夹板,胶合板
8. adhesive[əd'hiːsɪv; -zɪv]　*n.* & *adj.* 黏合剂;胶黏剂;黏着的;带黏性的
9. suspend[sə'spend]　*vt.* & *vi.* 延缓,推迟;使暂停;使悬浮;悬浮
10. longitudinal[,lɒndʒə'tudnl]　*adj.* 长度的,纵向的;经线的
11. stringer['strɪŋə]　*n.* 纵梁,纵桁
12. rigidity[rɪ'dʒɪdəti]　*n.* [物] 硬度,[力] 刚性
13. rivet['rɪvɪt]　*n.* & *vt.* 铆钉;铆接;固定
14. failure['feljə]　*n.* 失败;故障;失效;失败者;破产
15. assemble[ə'sembl]　*vt.* & *vi.* 集合,聚集;装配;收集;集合,聚集
16. auxiliary[ɔːg'zɪlɪəri]　*n.* & *adj.* 助动词;辅助物;附属机构;辅助的;副的;附加的
17. fairing['feərɪŋ]　*n.* [航] 整流罩
18. horizontal['hɒrə'zantl]　*adj.* & *n.* 水平的;地平线的;水平线,水平面;水平位置
19. contour['kantʊr]　*n.* & *vt.* 轮廓;等高线;周线;电路;概要;画轮廓;画等高线
20. stiffener['stɪfənə]　*n.* 加固物;[建] 加劲杆;刚性元件
21. substitute['sʌbstɪtut]　*n.* & *vt.* 代用品;代替者;替代;代替
22. lateral axis　[机] 横轴,[数] 横轴线
23. composite material　复合材料;聚合物复合材料
24. design criteria　设计准则;设计标准
25. failsafe['fel,sef]　*n.* 破损安全;失效保护
26. consist of　由……组成;由……构成;包括
27. extrusion[ɪk'struːʒn]　*n.* 挤出;推出;赶出;喷出
28. spar[spar]　*n.* 翼梁
29. trailing edge　(飞机的)机翼后缘
30. spanwise['spænwaiz]　*adj.* 顺翼展方向的
31. hinge[hɪndʒ]　*n.* & *v.* 铰链;关键,转折点;枢要,中枢;用铰链连接;依……为转移;给……安装铰链;(门等)装有蝶铰
32. aileron['eləran]　*n.* 副翼
33. rib[rɪb]　*n.* & *vt.* 肋骨;排骨;肋状物;翼肋;戏弄;装肋于
34. wing chord　[航] 翼弦
35. vulnerability[,vʌlnərə'bɪləti]　*n.* 易损性;弱点

36. countersunk screw　埋头螺丝;沉头螺钉

37. louver['luvə]　n.百叶窗;天窗;散热孔

38. cockpit['kakpIt]　n.驾驶员座舱;战场

39. aid with　用……来帮助

40. stiffen['stIfn]　vi.&vt.变硬;变猛烈;变黏;使变硬;使黏稠

41. honeycomb structure　[航]蜂窝状结构

42. nacelle[næ'sɛl]　n.气球吊篮;飞机的驾驶员室;飞机的引擎机舱

43. hydraulic[haI'drɔlIk]　adj.液压的;水力的;水力学的

44. vertical['vɜtIkl]　adj.&n.垂直的,直立的;垂直线,垂直面

45. elliptical[I'lIptIkl]　adj.椭圆的;省略的

46. firewall['faIəwɔl]　n.&vt.防火墙;用作防火墙

47. empennage[em'penIdʒ]　n.[航]尾翼,尾部

48. stabilizer['stebəlaIzə]　n.[助剂]稳定剂;稳定器;安定装置

49. tail cone　[航]尾锥;尾锥体;机身末端

50. symmetrically[sə'mɛtrIkli]　adv.对称地;平衡地;匀称地

51. compartment[kəm'partmənt]　n.&v.[建]隔间;区划;卧车上的小客房;分隔;划分

52. ventilation[ˌvɛntl'eʃən]　n.通风设备;空气流通

53. leakage['likIdʒ]　n.泄漏;渗漏物;漏出量

54. primary['praImɛri]　adj.&n.主要的;初级的;基本的;原色;最主要者

55. catastrophic[ˌkætə'strafIk]　adj.灾难的;悲惨的;灾难性的,毁灭性的

56. secondary structure　次要结构

57. inertial[I'nɜʃəl]　adj.惯性的;不活泼的

58. aerodynamic[ˌɛrodaI'næmIk]　adj.空气动力学的;[航]航空动力学的

59. emergency[I'mɜdʒənsi]　n.&adj.紧急情况;突发事件;非常时刻;紧急的;备用的

60. receptacle[rI'sɛptəkl]　n.[植]花托;容器;插座

61. fuselage['fjusəlaʒ]　n.[航]机身(飞机)

62. uniqueness[ju'niknIs]　n.独特性;独一无二;单值性

63. antenna[æn'tɛnə]　n.[电讯]天线

64. louver['luvə]　n.百叶窗;天窗;散热孔

Notes

(1) The skin of aircraft can also be made from a variety of materials, ranging from impregnated fabric to plywood, aluminum, or composites.

分析:该句子为被动语态;be made from 的意思是由"由……构成的,由……组成的",ranging from 的意为是"范围是从……到……"

翻译:飞机蒙皮同样也是由不同种类的材料构成的,其范围从浸渍(纤维)布到胶合板(多层板),铝,或者是复合材料。

(2) Their particular design for any given aircraft depends on a number of factors, such as

size，weight，use of the aircraft，desired speed in flight and at landing，and desired rate of climb.

分析：一般现在时的主动语态。整个句子的主语是"design"，主语的定语是"for any given aircraft"；谓语是一个词组"depends on"，为"取决于的意思"。

翻译：对于任何给定的飞机，其特定设计取决于很多因素，比如：(飞机)尺寸，重量，飞机的用途，设计的飞行和着陆速度以及设计的爬升率。

（3）Additionally，fail-safe spar web design exists. Fail-safe means that should one member of a complex structure fail，some other part of the structure assumes the load of the failed member and permits continued operation.

分析：fail-safe 的意思为破损安全，失效安全。

翻译：此外，大梁腹板采用破损安全设计思路。破损安全（失效安全）意味着一个具有复杂结构的单元失效，另外一些结构零件可以承受失效单元的载荷，并且允许其继续运行。

（4）One reason for this is the vulnerability of the wing tips to damage，especially during ground handling and taxiing.

分析：一般现在时的主动语态。句子主语为"One reason for this"。Vulnerability 的意思是"弱点，脆弱性"。

翻译：这样设计的原因之一是翼尖较易受损，尤其是在地面操作和滑行的过程中。

（5）A honeycomb structure is built up from a core material resembling a beehive's honeycomb which is laminated or sandwiched between thin outer skin sheets.

分析：一般现在时的被动语态。整个句子为 which 引导的非限制性定语从句，先行词是前面的整个句子。

翻译：蜂窝结构是用一种类似蜂窝形状的芯体构建的，蜂窝芯被夹在较薄的层合板之间形成层状结构或者三明治结构。

（6）On most single-engine aircraft，the engine and nacelle are at the forward end of the fuselage. On multiengine aircraft，engine nacelles are built into the wings or attached to the fuselage at the empennage (tail section).

分析：第一句话为一般现在时，主动语态；第二句话为一般现在时，被动语态。

翻译：在大部分单发飞机上，发动机和短舱位于机身的末端前部。对于多发飞机来说，发动机短舱吊挂在机翼上，或者安装在机身尾翼（尾段）上。

（7）The structure of the stabilizers is very similar to that which is used in wing construction.

分析：定语从句。"which is used in wing construction"是 that 的定语。which 引导的定语从句解释说明 that。

翻译：安定面的结构与机翼结构是非常相似的。

（8）All windows are designed to withstand cabin pressurization loads，and are designed with fail-safe features.

分析：一般现在时的被动语态。"cabin pressurization loads"的意思为客舱增压载荷。

翻译：所有风挡都被设计成能够承受客舱的增压载荷，并且具有破损安全结构特征。

（9）Primary structure is classified into two categories：Principal Structural Elements

(PSE) and Other Structure.

分析：一般现在时的被动语态。

翻译：主要结构被分成两大类：重要结构和其他主要结构。

(10)Fail-safe structure is ensured by the inner pane which is designed for 1.5 times the normal operating pressure at 70 degrees Fahrenheit.

分析：整个句子为一般现在时的被动语态，且为定语从句。Which 后面的内容跟 that 的内容一致，起到补充说明的作用。

翻译：内层玻璃为破损安全结构，其被设计成在 70 华氏度下能够承受 1.5 倍使用载荷。

Exercises

Ⅰ. Answer the following questions.

(1) How many units consists the airframe of a fixed-wing aircraft?

(2) How many types of fuselage construction?

(3) What components attach to the wing structure?

(4) In general, wing construction is based on what fundamental designs?

(5) Currently, what kind materials most manufactured aircraft wing spars?

(6) What means about fail-safe?

(7) Why the wing tip is often a removable unit, bolted to the outboard end of the wing panel?

(8) How many structural members consists the framework of a nacelle?

(9) What components consists the empennage?

(10) What is the role of conductive film on the windshield?

Ⅱ. Translating the following sentences into Chinese.

(1) The skin of aircraft can also be made from a variety of materials, ranging from impregnated fabric to plywood, aluminum, or composites.

(2) The semi-monocoque fuselage is constructed primarily of alloys of aluminum and magnesium, although steel and titanium are sometimes found in areas of high temperatures.

(3) Wings made entirely of carbon fiber or other composite materials exist, as well as wings made of a combination of materials for maximum strength to weight performance.

(4) Together, these members carry the loads caused by wing bending, with the caps providing a foundation for attaching the skin.

(5) A spar with failsafe construction is made in two sections. The top section consists of a cap riveted to the upper web plate. The lower section is a single extrusion consisting of the lower cap and web plate. These two sections are spliced together to form the spar. If either section of this type of spar breaks, the other section can still carry the load. This is the fail-safe feature.

(6) The wing tip is often a removable unit, bolted to the outboard end of the wing

panel.

(7) A louver on the top surface of the wingtip allows this warm air to be exhausted overboard.

(8) The engines are attached to the wing with struts (pylons). The strut is a pathway for all the pneumatic, electric, fuel, and hydraulic connections to the engine. The strut transfers loads from the engine and thrust reverser to the wing.

(9) A firewall is incorporated to isolate the engine compartment from the rest of the aircraft. This is basically a stainless steel or titanium bulkhead that contains a fire in the confines of the nacelle rather than letting it spread throughout the airframe.

(10) Flight control surfaces are usually similar in construction to one another and vary only in size, shape, and methods of attachment.

(11) The sealants that are used on the windows prevent moisture penetration, water entrapment, and provide aerodynamic flushness of the outer window pane and the window frame.

(12) Primary structure which contribute significantly to carrying flight, ground, and pressurization loads, and whose failure could result in the catastrophic failure of the airplane.

(13) Structure which carries only air or inertial loads generated on or within the secondary structure. Most secondary structures are important to the aerodynamic performance of the airplane.

Ⅲ. **Choose the right English meaning for following terms.**

(1)硬壳式

A. truss   B. monocoque   C. semi-monocoque

(2)纵梁

A. longeron   B. frame   C. rib

(3)龙骨梁

A. keel beam   B. frame   C. stringer

(4)桁条

A. bulkhead   B. frame   C. stringer

(5)机身站位

A. body station line   B. buttock line   C. water line

(6)吊架

A. nacelle   B. pylon   C. midspar

(7)应急门

A. Emergency Exits   B. Cargo Doors   C. Service Doors

(8)重要结构

A. primary structure   B. principal structural elements   C. secondary structure

Extensive Reading

Typical Airplane Fuselage Structures

The fuselage is a semi-monocoque structure consisting of skin reinforced by circumferential frames and longitudinal stringers. Major fuselage components are illustrated in Figure 6 – 17.

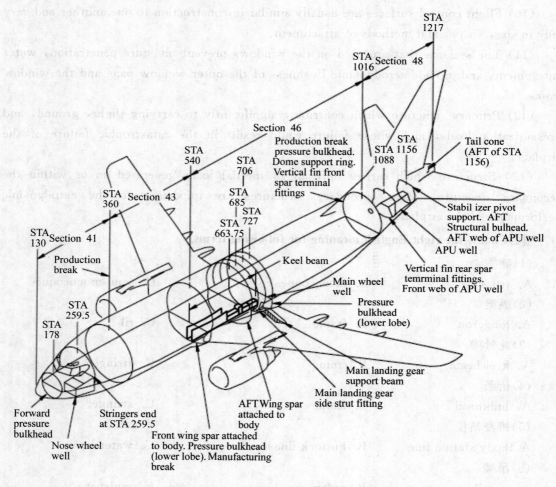

Figure 6 – 17   Major fuselage components

A typical section through the fuselage consists of an upper oval lobe and a lower oval lobe which intersect approximately at the floor level. At the intersection the fuselage is reinforced by transverse floor beams. Above this floor structure, which extends from the front pressure bulkhead at body station 178 to the rear pressure bulkhead at body station

1016, the upper lobe of the fuselage encloses the cabin and is basically a continuous shell, with cutouts in the skin for doors and windows. Below the floor the continuity of the lower lobe, which encloses the cargo compartments, is interrupted by several major structural features: the nose landing gear wheel well, the cavity for the center wing box and the main landing gear wheel well. Aft of the rear pressure bulkhead the floor is discontinued and this section of the fuselage, which tapers towards its aft end, supports the vertical fin, the horizontal stabilizer, a tail skid and contains a compartment with fireproof walls for the APU.

The various loads on the fuselage are due to combinations of flight, static, landing and pressure loads. Basically, the fuselage resembles a hollow tubular beam supported approximately at the center by the wing. This condition applies an overall bending load, to which may be added torsion applied by various flight maneuvers, and which is further complicated by pressurization loads acting on the whole body shell between stations 178 and 1016. Special design features maintain structural continuity between body stations 540 and 727, where the cavities for the center wing box and the main landing gear interrupts the lower half of the basically tubular fuselage. A keel beam connects the bottom of the fuselage frame at station 540 with the bottom of the frame at station 664 and passes below the center wing box.

The pressure wall over the center wing box area is formed by the upper surface of the wing itself, which also supports logitudinal cabin floor beams. A keel beam across the main landing gear wheel well connects the fuselage frames at stations 664 and 727. Over this area, the floor structure is plated and reinforced to form the pressure wall. A wing to fuselage connection exists by means of two six-flanged chords running between body stations 540 and 664, at buttock line 70.85 along the upper edges of the wing. The fuselage skin attaches to the upper flange of the six-flanged extrusions while the other five flanges are connections for the wing boxes to the center wing box. Main landing gear loads, other than those borne by the wing, are transmitted to the fuselage structure by the inboard ends of the landing gear support beams which are connected to fuselage structure at body station 706. All openings in the fuselage shell, such as those for doors and windows, are reinforced locally to maintain proper distribution of loads around the openings. The whole fuselage structure is designed on the fail-safe principle which provides alternate load paths to ensure that the entire fuselage is not jeopardized by the failure of any one member.

The fuselage is manufactured in four body sections, connected by production breaks or manufacturing breaks to form a complete integral structure. The forward three sections form the pressurized shell of the fuselage, and encloses all the passenger, crew, and cargo accommodations.

(1) Section 41 consists of that section of the fuselage between its extreme forward end and body station 360. Above the floor this section of the fuselage includes the control cabin, the forward entry door, and the forward galley door. Below the floor are the nose landing

gear wheel well, a lower nose compartment external access door, the forward airstairs and the electronic compartment. A bulkhead at body station 178 forms the forward pressure bulkhead of the whole fuselage, forward of which the nose radome is a nonstructural fairing. The frame at body station 360 is the production joint at which this section is attached to the second body section.

(2) Section 43 is that part of the fuselage between body stations 360 and 540. Above the floor it encloses the forward half of the passenger cabin. Below the floor it encloses the forward cargo compartment whose door is on the lower right side of the fuselage.

(3) Section 46 is that part of the fuselage between body station 540 and the rear pressure bulkhead at station 1016. The bulkhead at body station 540 is the joint at which this section is attached to section 43. Above the floor, section 46 encloses the aft half of the passenger cabin, including emergency exit hatches, the aft entry door, and the aft galley door. Below the floor it includes the cavity for the center wing box, the main landing gear wheel well and the aft cargo compartment, whose door is on the lower right side of the fuselage.

(4) Section 48 is not pressurized and extends aft from the rear pressure bulkhead at body station 1016. On top of section 48, at stations 1016 and 1088, the vertical fin attaches to four fittings, two front and two rear. A tail cone extends aft from station 1156. A compartment with fireproof walls in the lower part of the section, below the horizontal stabilizer, allows installation of the APU. The right rear torque box is sealed and sound-proofed to act as the APU air inlet duct. The horizontal stabilizer center section truss has its hinge joints by means of fittings attached to the bulkhead at station 1156. The front part of the center section truss protrudes through a cutout in the bulkhead at station 1088 and is moved up and down by a jackscrew unit fastened to the forward side of the station 1088 bulkhead. The left and right outboard sections of the horizontal stabilizer are cantilevered from the center section truss by means of fittings at the front and rear spars. An access and blowout door 3701 is located on the left side. Refer to Chapter 52, Doors, for maintenance practices regarding this door.

New Words/Phrases/Expression

1. frame[freɪm]　n.&v. 框架;结构;设计;建造
2. stringer['strɪŋə]　n. 纵梁,纵桁;上弦匠
3. lobe[ləʊb]　n.(脑、肺等的)叶;裂片;耳垂;波瓣
4. pressure bulkhead　气密框;气密隔板
5. continuity['kɒntə'nuəti]　n. 连续性;一连串;分镜头剧本
6. enclose[ɪn'kloz]　v. 围绕;装入;放入封套
7. interrupt['ɪntə'rʌpt]　v.&n. 中断;打断;插嘴;妨碍打扰
8. accommodation[ə,kɒmə'deʃən]　n. 住处,膳宿;调节;和解;预订铺位
9. extreme[ɪk'strim]　adj.&n. 极端的;极度的;极端;末端

10. rear[rɪə]　*adv. & adj. & n.* 向后;在后面;后方的;后面的;背面的;后面;屁股;后方部队

11. hatch[hætʃ]　*n.* [船][航]舱口

12. cavity['kævəti]　*n.* 腔;洞,凹处

13. vertical fin　垂尾,垂直安定面

14. tail cone　[航]尾锥;尾锥体;机身末端

15. sound-proof　*adj.* 隔音的

16. torque box　扭力盒

17. jackscrew['dʒæk,skru]　*n.* 螺旋起重机;起重螺旋;顶丝

18. cantilever['kæntɪlivə]　*n.* 悬臂

19. blowout['bloʊaʊt]　*n.* 爆裂;喷出;[电] 保险丝烧断

20. hollow['halo]　*adj. & n.* 空的;中空的,空腹的;凹的;虚伪的;洞;山谷;窟窿

21. tubular['tubjələ]　*adj.* 管状的

22. approximately[əˈpraksɪmətli]　*adv.* 大约,近似地;近于

23. bending load　弯曲负荷;弯曲荷载

24. bottom['batəm]　*n. & adj.* 底部;末端;底部的

25. reinforce[,riɪnˈfɔrs]　*vt. & vi. & n.* 加强,加固;强化;求援

26. flange[flændʒ]　*n. & vt.* [机] 法兰;给……装凸缘

27. borne[bɔrn]　*v.* 忍受;负荷

28. jeopardize['dʒɛpə,daɪz]　*vt.* 危害;使陷危地;使受困

29. integral['ɪntɪgrəl]　*adj. & n.* 完整的;整体的;必需的;积分;部分;完整

30. pressurize['prɛʃəraɪz]　*vt.* 密封;增压;使……加压,使……压入

31. galley['gæli]　*n.* [船] 船上的厨房

32. nonstructural　不作结构材料的

33. fireproof['faɪəpruːf]　*adj. & vt.* 防火的;耐火的;使耐火;使防水

34. maintenance['mentənəns]　*n.* 维护,维修;保持;生活费用

Exercises for Extensive Reading

Ⅰ. **Put the following sentences into Chinese.**

(1)A typical section through the fuselage consists of an upper oval lobe and a lower oval lobe which intersect approximately at the floor level.

(2)The various loads on the fuselage are due to combinations of flight, static, landing and pressure loads.

(3)A wing to fuselage connection exists by means of two six-flanged chords running between body stations 540 and 664, at buttock line 70. 85 along the upper edges of the wing.

(4)The fuselage is manufactured in four body sections, connected by production breaks or manufacturing breaks to form a complete integral structure.

(5)A bulkhead at body station 178 forms the forward pressure bulkhead of the whole fuselage, forward of which the nose radome is a nonstructural fairing.

(6) A compartment with fireproof walls in the lower part of the section, below the horizontal stabilizer, allows installation of the APU.

(7) The left and right outboard sections of the horizontal stabilizer are cantilevered from the center section truss by means of fittings at the front and rear spars.

Lesson 7 Structure Damage of Aircraft

INTENSIVE READING

Classification and Inspection of Damage

Inspecting Of Damage

When visually inspecting damage, remember that there may be other kinds of damage than that caused by impact from foreign objects or collision. A rough landing may overload one of the landing gear, causing it to become sprung; this would be classified as load damage. During inspection and "sizing up of the repair job", consider how far the damage caused by the sprung shock strut extends to supporting structural members.

A shock occurring at one end of a member will be transmitted throughout its length; therefore, inspect closely all rivets, bolts, and attaching structures along the complete member for any evidence of damage. Make a close examination for rivets that have partially failed and for holes which have been elongated.

Whether specific damage is suspected or not, an aircraft structure must occasionally be inspected for structural integrity. The following paragraphs provide general guidelines for this inspection.

When inspecting the structure of an aircraft, it is very important to watch for evidence of corrosion on the inside. This is most likely to occur in pockets and corners where moisture and salt spray may accumulate; therefore, drain holes must always be kept clean.

While an injury to the skin covering caused by impact with an object is plainly evident, a defect, such as distortion or failure of the substructure, may not be apparent until some evidence develops on the surface, such as canted, buckled or wrinkled covering, and loose rivets or working rivets. A working rivet is one that has movement under structural stress, but has not loosened to the extent that movement can be observed. This situation can sometimes be noted by a dark, greasy residue or deterioration of paint and primers around rivet heads. External indications of internal injury must be watched for and correctly interpreted. When found, an investigation of the substructure in the vicinity should be made and corrective action taken.

Warped wings are usually indicated by the presence of parallel skin wrinkles running diagonally across the wings and extending over a major area. This condition may develop from unusually violent maneuvers, extremely rough air, or extra hard landings. While there may be no actual rupture of any part of the structure, it may be distorted and weakened. Similar failures may also occur in fuselages. Small cracks in the skin covering may be caused by vibration and they are frequently found leading away from rivets.

Aluminum alloy surfaces having chipped protective coating, scratches, or worn spots that expose the surface of the metal should be recoated at once, as corrosion may develop rapidly. The same principle is applied to aluminum clad surfaces. Scratches, which penetrate the pure aluminum surface layer, permit corrosion to take place in the alloy beneath.

A simple visual inspection cannot accurately determine if suspected cracks in major structural members actually exist or the full extent of the visible cracks. Eddy current and ultrasonic inspection techniques are used to find hidden damage.

Another kind of damage to watch for is that caused by weathering or corrosion. This is known as corrosion damage. Corrosion damage of aluminum material is usually detected by the white crystalline deposits that form around loose rivets, scratches, or any portion of the structure that may be a natural spot for moisture to settle.

Classification of Damage

Damages may be grouped into four general classes. In many cases, the availability or lack of repair materials and time are the most important factors in determining whether a part should be repaired or replaced.

Negligible damage

Negligible damage consists of visually apparent, surface damage that do not affect the structural integrity of the component involved. Negligible damage may be left or may be corrected by a simple procedure without restricting flight. In both cases, some corrective action must be taken to keep the damage from spreading. Negligible or minor damage areas must be inspected frequently to ensure the damage does not spread. Permissible limits for negligible damage vary for different components of different aircraft and should be carefully researched on an individual basis. Failure to ensure that damages within the specified limit of negligible damage may result in insufficient structural strength of the affected support member for critical flight conditions.

Small dents, scratches, cracks, and holes that can be repaired by smoothing, sanding, stop drilling, or hammering out, or otherwise repaired without the use of additional materials, fall in this classification.

Damage Repairable by patching

Damage repairable by patching is any damage exceeding negligible damage limits that can be repaired by installing splice members to bridge the damaged portion of a structural part. The splice members are designed to span the damaged areas and to overlap the existing undamaged surrounding structure. The splice or patch material used in internal riveted and bolted repairs is normally the same type of material as the damaged part, but one gauge heavier. In a patch repair, filler plates of the same gauge and type of material as that in the damaged component may be used for bearing purposes or to return the damaged part to its original contour. Structural fasteners are applied to members and the surrounding structure to restore the original load-carrying characteristics of the damaged area. The use of patching depends on the extent of the damage and the accessibility of the component to be repaired.

Damage repairable by insertion

Damage must be repaired by insertion when the area is too large to be patched or the structure is arranged such that repair members would interfere with structural alignment (e. g. , in a hinge or bulkhead). In this type of repair, the damaged portion is removed from the structure and replaced by a member identical in material and shape. Splice connections at each end of the insertion member provide for load transfer to the original structure.

Damage Necessitating Replacement of Parts

Components must be replaced when their location or extent of damage makes repair impractical, when replacement is more economical than repair, or when the damaged part is relatively easy to replace. For example, replacing damaged castings, forgings, hinges, and small structural members, when available, is more practical than repairing them. Some highly stressed members must be replaced because repair would not restore an adequate margin of safety.

Replacement of an entire part is considered when one or more of the following conditions exist:

(1) When a complicated part has been extensively damage.

(2) When surrounding structure or inaccessibility makes repair impractical.

(3) When damaged part is relatively easy to replace.

(4) When forged or cast fitting are damaged beyond the negligible limits.

In structural repair manual, the term "damage" is defined as a cross-sectional area change or a permanent distortion of a structural member. Use the terms that follow:

(1) "Allowable Damage" is defined as damage that is permitted with no other flight restrictions.

(2) "Repairable Damage" is defined as damage that can be reworked or repaired.

(3) "Replacement of Damaged Parts" is defined as damage where the part must be

replaced.

You must decide what type of damage has occurred to a structural member or to a structural material.

Definition of Damage and Defects

Types of damage and defects which may be observed on parts of this assembly are defined as follows:

Abrasion: A damaged area that is the result of scuffing, rubbing, or other surface erosion. This type of damage is usually rough and has an irregular shape. The surface is worn away by either natural (rain or wind), mechanical (two parts that rub together), or man-made (oversanding, contact from tools and ground equipment, etc.). In a composite (fiber reinforced plastic), an abrasion does not go through the resin and into the fibers.

Adhesive Failure: Rupture of an adhesive bond such that the separation appears to be at the adhesive-adhered interface.

Broken Fibers: Fractured fibers in a fiber reinforced plastic caused by a gouge, puncture, or excessive bending or bearing load. Broken fiber damage in a composite structure can look similar to a crack in a metal structure.

Buckle: A mode of failure generally characterized by an unstable lateral material deflection that is a result of compressive action on the structural part. This type of buckling can occur in both metallic and nonmetallic materials. In advanced composites, buckling may take the form not only of conventional general instability and local instability, but also a micro instability of the individual fibers.

Burr: A small, thin section of metal extending beyond a regular surface, usually located at a corner or on the edge of a bore or hole.

Core Distortion: A condition that causes the shape of the core cells to change from a normal shape to a skewed shape.

Corrosion: Loss of metal from the surface by chemical or electrochemical action. The corrosion products generally are easily removed by mechanical means. Iron rust is an example of corrosion. Damage is the result of a complex electro-chemical action, and gives a cross-sectional area change. The depth of this damage must be determined by a cleanup or a removal operation.

Crack (Metal): A discontinuity (partial fracture or a full break) in the material that causes a significant cross-sectional area change. This damage usually has an irregular line and is often the result of fatigue in the material.

Crack (Composite Material): (a) Fractures in either the resin matrix or both the matrix and fibers. It is possible for a crack to go through some but not all of the plies. (b) A discontinuity (gap or lack of cohesion) which has a relatively large cross section in one direction and a small or negligible cross section when viewed in a direction perpendicular to the first.

Crazing: Region of ultrafine cracks which may extend in a network on or under the surface of a resin or plastic material can appear as a white band.

Crease: A damaged area that is depressed or folded back so that its boundaries are sharp or with well defined lines or ridges. Consider a crease in metal materials to be equal to a crack. In composites, you will see a break or line caused by a sharp fold (wrinkle).

Cut: Loss of metal, usually to an appreciable depth over a relatively long and narrow area, by mechanical means, as would occur with the use of a saw blade, chisel or sharp-edged stone striking a glancing blow.

Dent (Metal): A damaged area that is pushed in from its normal contour with no change in the cross-sectional area of the material. The edges of the damaged area are smooth. This damage is usually caused by a hit from a smoothly contoured object. The length of the dent is the longest distance from one end to the other end. The width of the dent is the second longest distance across the dent, measured at 90 degrees to the direction of the length.

Dent (Composite Material): A concave depression which does not break the fibers. A dent can cause a delamination or disbond.

Disbond: An area where there is a separation along the interface between two parts that are adhesively bonded. A disbond can be either a local separation or the complete loss of adhesion between a skin and the other laminate or between a skin and core.

Delamination: A type of disbond that occurs between adjacent plies of material. (can be local or over a wide area). If an allowable damage section or repair section does not give limits for disbonds, then use the limits specified for delaminations.

Edge Delamination: A separation of the skins or plies is from each other at the edge of a part.

Edge Void (Composite Material): An area where there is no adhesive along an edge after the cure.

Erosion: Destruction of metal or other material by the abrasive action of liquid or gas, usually accelerated by the presence of solid particles of matter in suspension and sometimes by corrosion. The eroded area will be rough and may be lined in the direction in which the foreign material moved relative to the surface.

Edge Erosion: Loss of paint, resin, and/or fibers along the outside surface or forward edge of a reinforced plastic (advanced composite) panel. Edge erosion is caused by the air, rain, snow, and ice that flow across the forward edge of the panel.

Exfoliation (Metal): A type of intergranular corrosion in metals, especially some aluminum alloys. Exfoliation occurs when corrosion at the grain boundaries below the surface pushes up the metallic grains on the surface, and most commonly on extruded sections.

Exfoliation (Composite): A surface defect on composite parts where the resin appears scaled or flaky.

Fatigue: The failure or decay of mechanical properties after repeated applications of stress. The result will be a fracture under repeated or fluctuating stresses (loads) having a maximum value less than the tensile strength of the material. Fatigue fractures will begin as microscopic cracks that will grow under the action of the fluctuating (cyclic) stress (load).

Fiber or Resin Pullout: Fiber or resin pullout occurs when small pieces of resin or composite fibers are pulled away from the matrix during a cut or drill procedure.

Galling: Breakdown (or build-up) of metal surfaces due to excessive friction between two parts having relative motion. Particles of the softer metal are torn loose and "welded" to the harder.

Galvanic Corrosion: Corrosion associated with the current of a galvanic cell made up of dissimilar electrodes.

Gouge: A damaged area where the result is a cross-sectional changeis caused by a sharp object and gives a continuous, sharp or smooth groove in the material.

Heat Damage: Loss of strength and/or loss of material in an area of structure are/is caused by exposure to very high temperature for a short period of time, or by a moderately high temperature for a long period of time.

Hole: A puncture or cut-out that is fully surrounded by undamaged material. Other types of damage can be removed by making an oversized hole or a hole that has an irregular shape, if this hole stays in the allowable damage limits or can be repaired with an approved procedure.

Inclusion: Presence of foreign or extraneous material wholly within a portion of metal. Such material is introduced during the manufacture of rod, bar or tubing by rolling or forging.

Lightning Strike Damage: Loss of strength and/or loss of material in an area of structure that was caused by an attachment of a lightning bolt. Damage can be seen as discoloration, pits, holes, and/or melted material.

Matrix Imperfections: (Micro-cracks, porosity, blister, etc.) These can occur in the resin matrix during the cure cycle on the resin matrix-fiber interface, or in the resin matrix parallel to the fibers.

Nick: Local break or notch on edge, usually displacement of metal rather than loss. You can consider a series of nicks in a line pattern to be equal to a gouge.

Pit: A form of corrosion that occurs under particular conditions which result in small craters instead of a wide area. A pit is sharp, localized breakdown (small, deep cavity) of metal surface, usually with defined edges.

Ply Wrinkle: An out-of-plane distortion of fibers in a cured composite part.

Score: Deeper (than scratch) tear or break in metal surface from contact under pressure.

Scratch: Slight tear or break in metal surface from light, momentary contact by foreign material.

Stress Corrosion: Corrosion that occurs in a static or dynamically loaded area of a metal part in a corrosive environment. Stress corrosion can occur in an environment that will not necessarily cause corrosion in a part that is not loaded.

Void: An area where the adhesive is missing, or a delamination, or an area where there the mating surfaces do not touch, or an empty space in a cured resin-fiber system.

Damage Tolerance Definitions and Repair Classification

What's the meaning of the Damage Tolerance? In brief, the ability of structure to sustain anticipated loads in the presence of damage, such as fatigue cracks, until it is detected through inspection or malfunction and repaired.

There are two classifications of repairs in the aircraft maintenance project:

1. Repairs that have been evaluated and analyzed for damage tolerance capability and are classified as Category A, B or C repairs.

2. Repairs which are not critical for the damage tolerance capability of the airplane are classified as permanent, interim, or time-limited, based on the expected durability of the repair. If a repair is not identified as an interim or time-limited repair it is a permanent repair.

The definitions of the different categories of damage tolerant repairs are as follows:

1. Category A Repair: A permanent repair for which the inspections given in the Maintenance Planning Data (MPD) document and any applicable Airworthiness Directives are sufficient and no other actions are necessary.

2. Category B Repair: A permanent repair for which additional inspections are necessary at the specified threshold and repeat intervals.

3. Category C Repair: A time-limited repair which must be replaced or reworked within a specified time limit. Also supplemental inspections can be necessary at a specified threshold and repeat interval.

The following definitions will be used for repairs which are not critical for damage tolerance. They are classified as permanent, interim, or time-limited, based on the expected durability of the repair. If a repair is not identified as an interim or time-limited repair, or classified as Category A, B or C, then it is a permanent repair.

1. Permanent Repair: A repair where no action is necessary except the operator's normal maintenance.

2. Interim Repair: A repair that has the necessary structural strength and could stay on the aircraft indefinitely. The repair must be inspected at specified intervals and replaced if deterioration is detected or damage is found.

3. Time-Limited Repair: A repair that has the necessary structural strength but does not have sufficient durability. The repair must be replaced after a specified time, usually given as a number of flight cycles, flight hours or a calendar time.

New Words/Phrases/Expression

1. classification[ˌklæsɪfɪˈkeʃən]　*n.* 分类；类别，等级

2. inspection[ɪnˈspɛkʃən]　*n.* 视察，检查

3. visually[ˈvɪʒʊəli]　*adv.* 形象化地；外表上；看得见地

4. collision[kəˈlɪʒən]　*n.* 碰撞；冲突

5. overload[əʊvəˈləʊd]　*vt.&n.* 超载，超过负荷；超载；负荷过多

6. elongate[ˈiːlɒŋɡeɪt]　*vt.&adj.&vi.* 拉长；使延长；使伸长；拉长；延长；伸长；伸长的；延长的

7. structural integrity　结构完整性

8. general guideline　一般准则

9. moisture[ˈmɔɪstʃə]　*n.* 水分；湿度；潮湿；降雨量

10. spray[spreɪ]　*n.&vi.&vt.* 喷雾；喷雾器；水沫；喷射喷

11. accumulate[əˈkjumjəlet]　*v.* 累积；积聚；积攒

12. injury[ˈɪndʒəri]　*n.* 伤害，损害；受伤处

13. plainly[ˈpleɪnli]　*adv.* 明显地；清楚地；简单地；坦率地

14. defect[ˈdɪfɛkt]　*n.* 缺点，缺陷；不足之处

15. distortion[dɪsˈtɔrʃən]　*n.* 变形；[物] 失真；扭曲；曲解

16. substructure[ˈsʌbstrʌktʃə]　*n.* 基础；子结构；底部构造

17. cant[kænt]　*n.&v.* 斜面；伪善之言；黑话；角落；把……棱角去掉；使……倾斜；甩掉

18. buckle[ˈbʌk(ə)l]　*v.&n.* 扣住；变弯曲；使弯曲；皮带扣，带扣

19. wrinkle[ˈrɪŋkl]　*n.&v.* 皱纹；起皱；使起皱纹

20. greasy[ˈɡrisi]　*adj.* 油腻的；含脂肪多的；谄媚的

21. residue[ˈrɛzɪdu]　*n.* 残渣；剩余；滤渣

22. deterioration[dɪˌtɪrɪəˈreʃən]　*n.* 恶化；退化；堕落

23. vicinity[vəˈsɪnəti]　*n.* 邻近，附近；近处

24. parallel[ˈpærəlɛl]　*n.&vt.&adj.* 平行线；对比；使……与……平行；平行的；类似的，相同的

25. violent[ˈvaɪələnt]　*adj.* 暴力的；猛烈的

26. maneuver[məˈnuvə]　*v.&n.* 机动，演习

27. rupture[ˈrʌptʃə]　*n.&v.* 破裂；决裂；疝气；破裂；发疝气；使破裂；断绝；发生疝

28. vibration[vaɪˈbreʃən]　*n.* 振动；犹豫；心灵感应

29. protective coating　保护涂料；保护层；保护涂层

30. crystalline[ˈkrɪstəlaɪn]　*adj.* 透明的；水晶般的；水晶制的

31. negligible[ˈnɛɡlɪdʒəbl]　*adj.* 微不足道的，可以忽略的

32. permissible limit　容许极限，允许的限度

33. contour[ˈkantʊr]　*n.&vt.* 轮廓；等高线；周线；电路；概要；画轮廓；画等高线

34. accessibility[ækˌsɛsəˈbɪləti]　*n.* 易接近；可亲；可以得到

35. insertion[ɪnˈsɜ˞ʃɪnɜl] n. 插入;嵌入;插入物

36. alignment[əˈlaɪnmənt] n. 队列,成直线;校准;结盟

37. identical[aɪˈdentɪkl] adj. & n. 同一的;完全相同的;完全相同的事物

38. cross-sectional area 横断面积;断面面积

39. permanent distortion 永久变形;残留变形

40. unstable[ʌnˈstebl] adj. 不稳定的;动荡的;易变的

41. deflection[dɪˈflekʃən] n. 偏向;挠曲;偏差

42. nonmetallic material 非金属材料

43. electrochemical action 电化学作用

44. crazing[ˈkreɪzɪŋ] n. 银纹;破裂;龟裂;细裂纹

45. ultrafine[ˌʌltrəˈfain] adj. 非常细微的

46. crease[kris] n. ,v. 折痕;折缝;起皱弄皱;使起折痕

47. erosion[ɪˈroʒən] n. 磨蚀,侵蚀,腐蚀

48. abrasive action 磨损作用

49. exfoliation[eks,folɪˈeɪʃən] n. 剥落;剥落物;表皮脱落

50. fluctuating stress 脉动应力;交变载荷

51. galvanic corrosion 接触腐蚀;电偶腐蚀;电化学腐蚀

52. gouge[gaʊdʒ] v. 用半圆凿子挖;欺骗 n. 沟;圆凿;以圆凿刨

53. moderately[ˈmadərətli] adv. 适度地;中庸地;有节制地

54. puncture[ˈpʌŋktʃə] v. & n. 刺穿;揭穿;削弱;被刺穿;被戳破;穿刺;刺痕

55. inclusion[ɪnˈkluʒn] n. 包含;内含物

56. extraneous material 外来杂质

57. lightning strike damage 雷击损伤

58. discoloration[ˌdɪs,kʌləˈreɪʃən] n. 变色;污点

59. momentary contact 瞬间接触;瞬时接触

60. stress corrosion [力] 应力腐蚀

61. damage tolerance 损伤容限

62. anticipate[ænˈtɪsəˈpeɪt] vt. 预期,期望;占先,抢先;提前使用

63. fatigue crack [力] 疲劳裂纹

64. malfunction[ˌmælˈfʌŋkʃən] vi. 发生故障;不起作用 n. 故障;失灵;疾病

65. evaluate[ɪˈvæljʊeɪt] v. 评价;估价;求……的值;评价;估价

66. durability[ˌdjʊrəˈbɪləti] n. 耐久性;坚固;耐用年限

67. threshold[ˈθreʃhold] n. 入口;门槛;开始;极限;临界值

68. interim[ˈɪntərɪm] adj. & n. 临时的,暂时的;中间的;间歇的;过渡时期,中间时期;暂定

Notes

(1) When visually inspecting damage, remember that there may be other kinds of

— 115 —

damage than that caused by impact from foreign objects or collision.

分析：省略了主语的宾语从句。When 引导的是时间状语；that 后面接"there be"句型，是整个句子的宾语。

翻译：当目视检查损伤的时候，记得除由外来物和撞击引起的损伤外，可能还存在其他类型的损伤。

（2）While an injury to the skin covering caused by impact with an object is plainly evident，a defect，such as distortion or failure of the substructure，may not be apparent until some evidence develops on the surface，such as canted，buckled or wrinkled covering，and loose rivets or working rivets.

分析：整个句子为"not ... until ..."的句式结构，意思为"直到……才会……"

翻译：然而，由外来物冲击引起的蒙皮损伤是非常明显的，而一种缺陷，比如说底层结构的扭曲变形或者失效可能只有当其发展为表面的诸如倾斜，塌陷或皱褶蒙皮和铆钉松落才会明显。

（3）Aluminum alloy surfaces having chipped protective coating，scratches，or worn spots that expose the surface of the metal should be recoated at once，as corrosion may develop rapidly.

分析：条件状语从句，翻译为"如果……就……"

翻译：如果铝合金表面具有保护涂层破损，划伤，或者磨损点，那么金属的表面应该及时重新添加涂层，因为腐蚀的发展速度会非常迅速。

（4）Negligible damage consists of visually apparent，surface damage that do not affect the structural integrity of the component involved.

分析：damage 的具体内容为"that do not affect the structural integrity of the component involved"。

翻译：可忽略的损伤是由那些容易看到的表面损伤组成的，这些损伤不会影响所涉及的部件的结构完整性。

（5）Damage must be repaired by insertion when the area is too large to be patched or the structure is arranged such that repair members would interfere with structural alignment.

分析：条件状语从句。

翻译：当损伤面积大到不能用补丁修理，或者该结构中存在某修理单元会破坏结构完整性时，这类损伤必须通过嵌入件来修理。

（6）"Replacement of Damaged Parts" is defined as damage where the part must be replaced.

分析："... be defined as ..."翻译为"被定义为……"

翻译："易损件更换"被定义为损伤处的零件必须被更换。

（7）A mode of failure generally characterized by an unstable lateral material deflection that is a result of compressive action on the structural part.

分析："generally characterized by an unstable lateral material deflection"为"A mode of failure"的定语，也就是过去分词做主语的定语。整个句子为主语从句。

翻译：一种通常被描述为不稳定的横向材料扭转（扭曲）变形的失效模式，是由作用在结构零件上的压应力造成。

Exercises

Ⅰ. Answer the following questions.

(1) According to the text, how many types of damage and defects are defined?

(2) How many general classes may damages be grouped into? What are they?

(3) When is replacement of an entire part considered?

(4) Which damage is classified as damage repairable by insertion?

(5) Where does a burr usually locate?

(6) What may it be called if there is a physical separation of two adjacent portions of metal which can be evidenced by a fine or thin line across the surface?

Ⅱ. Translating the following sentences into Chinese.

(1) Whether specific damage is suspected or not, an aircraft structure must occasionally be inspected for structural integrity.

(2) Warped wings are usually indicated by the presence of parallel skin wrinkles running diagonally across the wings and extending over a major area.

(3) A simple visual inspection cannot accurately determine if suspected cracks in major structural members actually exist or the full extent of the visible cracks.

(4) Negligible or minor damage areas must be inspected frequently to ensure the damage does not spread.

(5) Structural fasteners are applied to members and the surrounding structure to restore the original load-carrying characteristics of the damaged area.

(6) In advanced composites, buckling may take the form not only of conventional general instability and local instability, but also a micro instability of the individual fibers.

(7) The depth of this damage must be determined by a cleanup or a removal operation.

(8) The failure or decay of mechanical properties after repeated applications of stress. The result will be a fracture under repeated or fluctuating stresses (loads) having a maximum value less than the tensile strength of the material. Fatigue fractures will begin as microscopic cracks that will grow under the action of the fluctuating (cyclic) stress (load).

(9) Other types of damage can be removed by making an oversized hole or a hole that has an irregular shape, if this hole stays in the allowable damage limits or can be repaired with an approved procedure.

(10) Repairs which are not critical for the damage tolerance capability of the airplane are classified as permanent, interim, or time-limited, based on the expected durability of the repair.

(11) Time-Limited Repair: A repair that has the necessary structural strength but does not have sufficient durability. The repair must be replaced after a specified time, usually

given as a number of flight cycles, flight hours or a calendar time.

Ⅲ. **Choose the right English meaning for following terms.**

(1)允许损伤

    A. Allowable Damage   B. Repairable Damage   C. Replacement of Damaged Parts

(2)磨损

    A. buckle   B. abrasion   C. burr

(3)裂纹

    A. crazing   B. crease   C. crack

(4)分层

    A. delamination   B. disbond   C. cut

(5)凹坑

    A. dent   B. disbond   C. exfoliation

(6)应力腐蚀

    A. corrosion   B. stress corrosion   C. erosion

(7)电化学腐蚀

    A. corrosion   B. stress corrosion   C. Galvanic Corrosion

(8)永久性修理

    A. Interim Repair   B. Permanent Repair   C. Time-Limited Repair

EXTENSIVE READING

Corrosion

Many aircraft structures are made of metal, and the most insidious form of damage to those structures is corrosion. From the moment the metal is manufactured, it must be protected from the deleterious effects of the environment that surrounds it. This protection can be the introduction of certain elements into the base metal, creating a corrosion resistant alloy, or the addition of a surface coating of a chemical conversion coating, metal or paint. While in use, additional moisture barriers, such as viscous lubricants and protectants may be added to the surface.

The introduction of airframes built primarily of composite components has not eliminated the need for careful monitoring of aircraft with regard to corrosion. While the airframe itself may not be subject to corrosion, the use of metal components and accessories within the airframe means the aircraft maintenance technician must be on the alert for the evidence of corrosion when inspecting any aircraft.

Water or water vapor containing salt combines with oxygen in the atmosphere to produce the main source of corrosion in aircraft. Aircraft operating in a marine environment, or in areas where the atmosphere contains industrial fumes that are corrosive, are particularly susceptible to corrosive attacks.

The appearance of corrosion varies with the metal. On the surface of aluminum alloys and magnesium, it appears as pitting and etching, and is often combined with a gray or white powdery deposit. On copper and copper alloys, the corrosion forms a greenish film; on steel, a reddish corrosion byproduct commonly referred to as rust. When the gray, white, green, or reddish deposits are removed, each of the surfaces may appear etched and pitted, depending upon the length of exposure and severity of attack. If these surface pits are not too deep, they may not significantly alter the strength of the metal; however, the pits may become sites for crack development, particularly if the part is highly stressed. Some types of corrosion burrow between the inside of surface coatings and the metal surface, and can spread until the part fails.

Types of Corrosion

There are two general classifications of corrosion that cover most of the specific forms: direct chemical attack and electrochemical attack. In both types of corrosion, the metal is converted into a metallic compound such as an oxide, hydroxide, or sulfate. The corrosion process always involves two simultaneous changes: The metal that is attacked or oxidized suffers what may be called anodic change, and the corrosive agent is reduced and may be considered as undergoing cathodic change.

Direct Chemical Attack

Direct chemical attack, or pure chemical corrosion, is an attack resulting from a direct exposure of a bare surface to caustic liquid or gaseous agents. Unlike electrochemical attack where the anodic and cathodic changes may be taking place a measurable distance apart, the changes in direct chemical attack are occurring simultaneously at the same point. The most common agents causing direct chemical attack on aircraft are: (1) spilled battery acid or fumes from batteries; (2) residual flux deposits resulting from inadequately cleaned, welded, brazed, or soldered joints; and (3) entrapped caustic cleaning solutions.

Electrochemical Attack

The reaction in electrochemical attack requires a medium, usually water, which is capable of conducting a tiny current of electricity. When a metal comes in contact with a corrosive agent and is also connected by a liquid or gaseous path through which electrons may flow, corrosion begins as the metal decays by oxidation(Figure 7 – 1). During the attack, the quantity of corrosive agent is reduced and, if not renewed or removed, may completely react with the metal, becoming neutralized.

The conditions for these corrosion reactions are the presence of a conductive fluid and metals having a difference in potential. If, by regular cleaning and surface refinishing, the medium is removed and the minute electrical circuit eliminated, corrosion cannot occur. This is the basis for effective corrosion control. The electrochemical attack is responsible for most forms of corrosion on aircraft structure and component parts.

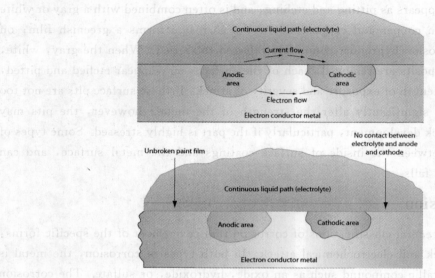

Figure 7 - 1   Electrochemical attack

Forms of Corrosion

There are many forms of corrosion. The form of corrosion depends on the metal involved, its size and shape, its specific function, atmospheric conditions, and the corrosion producing agents present. Those described in this section are the more common forms found on airframe structures.

Surface Corrosion

Surface corrosion appears as a general roughening, etching, or pitting of the surface of a metal, frequently accompanied by a powdery deposit of corrosion products. Surface corrosion may be caused by either direct chemical or electrochemical attack(Figure 7 - 2).

Filiform corrosion gives the appearance of a series of small worms under the paint surface. It is often seen on surfaces that have been improperly chemically treated prior to painting (Figure 7 - 3).

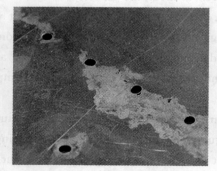

Figure 7 - 2   Surface corrosion

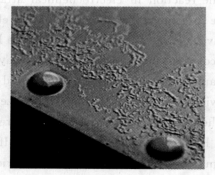

Figure 7 - 3   Filiform corrosion

Dissimilar Metal Corrosion

The contamination of a metal's surface by mechanical means can induce dissimilar metal corrosion. The improper use of steel cleaning products, such as steel wool or a steel wire brush on aluminum or magnesium, can force small pieces of steel into the metal being cleaned, which will then further corrode and ruin the adjoining surface. Carefully monitor the use of nonwoven abrasive pads, so that pads used on one type of metal are not used again on a different metal surface (Figure 7 – 4).

Figure 7 – 4   Dissimilar metal corrosion

Intergranular Corrosion

This type of corrosion is an attack along the grain boundaries of an alloy and commonly results from a lack of uniformity in the alloy structure. Aluminum alloys and some stainless steels are particularly susceptible to this form of electrochemical attack (Figure 7 – 5). The lack of uniformity is caused by changes that occur in the alloy during heating and cooling during the material's manufacturing process. Intergranular corrosion may exist without visible surface evidence. Very severe intergranular corrosion may sometimes cause the surface of a metal to "exfoliate" (Figure 7 – 6). This type of corrosion is difficult to detect in its initial stage. Extruded components such as spars can be subject to this type of corrosion. Ultrasonic and eddy current inspection methods are being used with a great deal of success.

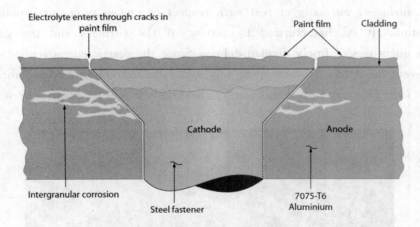

Figure 7 – 5   Intergranular corrosion of 7075 – T6 aluminum adjacent to steel fastener

Figure 7 – 6　Exfoliation

Stress Corrosion

Stress corrosion occurs as the result of the combined effect of sustained tensile stresses and a corrosive environment. Stress corrosion cracking is found in most metal systems; however, it is particularly characteristic of aluminum, copper, certain stainless steels, and high strength alloy steels (over 240,000 psi). Aluminum alloy bell cranks with pressed in bushings, landing gear shock struts with pipe thread type grease fittings, clevis pin joints, shrink fits, and overstressed tubing B-nuts are examples of parts which are susceptible to stress corrosion cracking.

Fretting Corrosion

Fretting corrosion is a particularly damaging form of corrosive attack that occurs when two mating surfaces, normally at rest with respect to one another, are subject to slight relative motion. It is characterized by pitting of the surfaces and the generation of considerable quantities of finely divided debris. Since the restricted movements of the two surfaces prevent the debris from escaping very easily, an extremely localized abrasion occurs (Figure 7 – 7). This type of corrosion (on bearing surfaces) has also been called false brinelling.

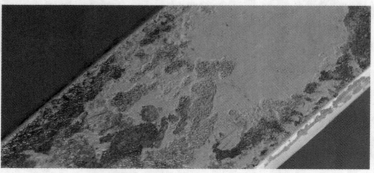

Figure 7 – 7　Fretting corrosion

Factors Affecting Corrosion

Many factors affect the type, speed, cause, and seriousness of metal corrosion. Some of these factors can be controlled and some cannot.

Climate

The environmental conditions under which an aircraft is maintained and operated greatly affect corrosion characteristics. In a predominately marine environment (with exposure to sea water and salt air), moisture-laden air is considerably more detrimental to an aircraft than it would be if all operations were conducted in a dry climate. Temperature considerations are important because the speed of electrochemical attack is increased in a hot, moist climate.

Foreign Material

Among the controllable factors which affect the onset and spread of corrosive attack is foreign material that adheres to the metal surfaces. Such foreign material includes:

1. Soil and atmospheric dust.
2. Oil, grease, and engine exhaust residues.
3. Salt water and salt moisture condensation.
4. Spilled battery acids and caustic cleaning solutions.
5. Welding and brazing flux residues.

It is important that aircraft be kept clean. How often and to what extent an aircraft should be cleaned depends on several factors, including geographic location, model of aircraft, and type of operation.

Preventive Maintenance

Much has been done to improve the corrosion resistance of aircraft: improvements in materials, surface treatments, insulation, and in particular, modern protective finishes. All of these have been aimed at reducing the overall maintenance effort, as well as improving reliability. In spite of these improvements, corrosion and its control is a very real problem that requires continuous preventive maintenance. Corrosion preventive maintenance includes the following specific functions:

1. Adequate cleaning.
2. Thorough periodic lubrication.
3. Detailed inspection for corrosion and failure of protective systems.
4. Prompt treatment of corrosion and touchup of damaged paint areas.
5. Keeping drain holes free of obstructions.
6. Daily draining of fuel cell sumps.
7. Daily wipe down of exposed critical areas.

8. Sealing of aircraft against water during foul weather and proper ventilation on warm, sunny days.

9. Maximum use of protective covers on parked aircraft.

Inspection

Inspection for corrosion is a continuing problem and should be handled on a daily basis. Overemphasizing a particular corrosion problem when it is discovered and then forgetting about corrosion until the next crisis is an unsafe, costly, and troublesome practice. Most scheduled maintenance checklists are complete enough to cover all parts of the aircraft or engine, and no part of the aircraft should go uninspected.

Corrosion Prone Areas

Discussed briefly in this section are most of the trouble areas common to all aircraft. However, this coverage is not necessarily complete and may be amplified and expanded to cover the special characteristics of the particular aircraft model involved by referring to the applicable maintenance manual.

Exhaust Trail Areas

Both jet and reciprocating engine exhaust deposits are very corrosive and give particular trouble where gaps, seams, hinges, and fairings are located downstream from the exhaust pipes or nozzles. Deposits may be trapped and not reached by normal cleaning methods. Pay special attention to areas around rivet heads and in skin lap joints and other crevices. Remove and inspect fairings and access plates in the exhaust areas. Do not overlook exhaust deposit buildup in remote areas, such as the empennage surfaces.

Battery Compartments and Battery Vent Openings

Despite improvements in protective paint finishes and in methods of sealing and venting, battery compartments continue to be corrosion prone areas. Fumes from overheated electrolyte are difficult to contain and will spread to adjacent cavities and cause a rapid corrosive attack on all unprotected metal surfaces. Battery vent openings on the aircraft skin should be included in the battery compartment inspection and maintenance procedure. Regular cleaning and neutralization of acid deposits will minimize corrosion from this cause.

Bilge Areas

These are natural sumps for waste hydraulic fluids, water, dirt, and odds and ends of debris. Residual oil quite often masks small quantities of water that settle to the bottom and set up a hidden chemical cell.

Inspection procedures should include particular attention paid to areas located under galleys and lavatories and to human waste disposal openings on the aircraft exteriors.

Human waste products and the chemicals used in lavatories are very corrosive to common aircraft metals. Clean these areas frequently and keep the paint touched up.

Wheel Well and Landing Gear

More than any other area on the aircraft, this area probably receives more punishment due to mud, water, salt, gravel, and other flying debris.

Because of the many complicated shapes, assemblies, and fittings, complete area paint film coverage is difficult to attain and maintain. A partially applied preservative tends to mask corrosion rather than prevent it. Due to heat generated by braking action, preservatives cannot be used on some main landing gear wheels.

Water Entrapment Areas

Design specifications require that aircraft have drains installed in all areas where water may collect. Daily inspection of low point drains should be a standard requirement. If this inspection is neglected, the drains may become ineffective because of accumulated debris, grease, or sealants.

Engine Frontal Areas and Cooling Air Vents

These areas are being constantly abraded with airborne dirt and dust, bits of gravel from runways, and rain erosion, which tends to remove the protective finish. Inspection of these areas should include all sections in the cooling air path, with special attention to places where salt deposits may be built up during marine operations.

Wing Flap and Spoiler Recesses

Dirt and water may collect in flap and spoiler recesses and go unnoticed because they are normally retracted. For this reason, these recesses are potential corrosion problem areas. Inspect these areas with the spoilers and/or flaps in the fully deployed position.

External Skin Areas

External aircraft surfaces are readily visible and accessible for inspection and maintenance. Even here, certain types of configurations or combinations of materials become troublesome under certain operating conditions and require special attention.

Trimming, drilling, and riveting destroy some of the original surface treatment, which is never completely restored by touchup procedures. Any inspection for corrosion should include all skin surfaces with special attention to edges, areas around fasteners, and cracked, chipped, or missing paint.

Piano-type hinges are prime spots for corrosion due to the dissimilar metal contact between the steel pin and aluminum hinge. They are also natural traps for dirt, salt, and moisture. Inspection of hinges should include lubrication and actuation through several cycles to ensure complete lubricant penetration. Use water-displacing lubricants when servicing piano hinges.

Corrosion of metal skins joined by spot welding is the result of the entrance and entrapment of corrosive agents between the layers of metal. This type of corrosion is evidenced by corrosion products appearing at the crevices through which the corrosive agents enter. More advanced corrosive attack causes skin buckling and eventual spot weld fracture. Skin buckling in its early stages may be detected by sighting along spot welded seams or by using a straightedge.

Miscellaneous Trouble Areas

All control cables, whether plain carbon steel or corrosion resistant steel, should be inspected to determine their condition at each inspection period. In this process, inspect cables for corrosion by random cleaning of short sections with solvent soaked cloths. If external corrosion is evident, relieve tension and check the cable for internal corrosion. Replace cables that have internal corrosion. Remove light external corrosion with a nonwoven abrasive pad lightly soaked in oil or, alternatively, a steel wire brush. When corrosion products have been removed, recoat the cable with preservative.

New Words/Phrases/Expression

1. corrosion[kə'rəuʒən] n.腐蚀;腐蚀产生的物质;衰败
2. insidious[ɪn'sɪdɪəs] adj.阴险的;隐伏的;暗中为害的;狡猾的
3. deleterious['dɛlə'tɪrɪəs] adj.有毒的,有害的
4. viscous['vɪskəs] adj.黏性的;黏的
5. lubricant['lubrɪkənt] n.,adj.润滑剂;润滑油;润滑的
6. protectant[prə'tɛktənt] n.保护剂;杀虫剂
7. marine environment [环境]海洋环境
8. susceptible[sə'sɛptəbl] adj.&n.易受影响的;易感动的;容许……的;易得病的人
9. powdery['paudəri] adj.粉的;粉状的;布满粉状物的
10. burrow['bɜ'rəu] v.&n.探索,寻找;挖掘,挖出;(兔、狐等的)洞穴,地道
11. metallic compound 金属化合物
12. oxide['aksaɪd] n.[化学]氧化物
13. hydroxide[haɪ'drɑksaɪd] n.[无化]氢氧化物;羟化物
14. sulfate['sʌlˌfeɪt] n.[无化]硫酸盐 v.使成硫酸盐;用硫酸处理;硫酸盐化
16. anodic[æn'ɑdɪk] adj.[电]阳极的;上升的
17. caustic['kɔstɪk] adj.&n.[化学]腐蚀性的;[助剂]腐蚀剂;苛性钠
18. neutralize['nʊtrə'laɪz] vt.&vi.抵销;使……中和;使……无效;使……中立;中和;中立化;变无效
19. surface corrosion 表面腐蚀

— 126 —

20. filiform corrosion　［涂料］丝状腐蚀

21. bell crank　曲拐；曲柄；直角杠杆；摇臂

22. false brinelling　摩擦腐蚀压痕

23. predominately[prɪˈdɒmɪnətlɪ]　*adv.* 占优势地；有影响力地；更大量地

24. geographic location　地理定位；地理位置

25. preventive maintenance　预防性维修；定期检修

26. protective finish　表面处理，［涂料］保护涂层

27. ventilation[ˌventɪˈleɪʃn]　*n.* 通风设备；空气流通

28. overemphasize[ˌəʊvərˈemfəsaɪz]　*v.* 过分强调

29. scheduled maintenance checklist　定期维护检查单；计划维护清单

30. seam[siːm]　*n.* & *v.* 缝；接缝；缝合；接合；使留下伤痕；裂开；产生裂缝

31. electrolyte[ɪˈlektrəlaɪt]　*n.* 电解液，电解质；电解

32. debris[dəˈbri]　*n.* 碎片，残骸

33. water-displacing lubricant　排水润滑剂

Exercises

Ⅰ. **Put the following sentences into Chinese.**

（1）While the airframe itself may not be subject to corrosion, the use of metal components and accessories within the airframe means the aircraft maintenance technician must be on the alert for the evidence of corrosion when inspecting any aircraft.

（2）The corrosion process always involves two simultaneous changes：The metal that is attacked or oxidized suffers what may be called anodic change, and the corrosive agent is reduced and may be considered as undergoing cathodic change.

（3）The form of corrosion depends on the metal involved, its size and shape, its specific function, atmospheric conditions, and the corrosion producing agents present.

（4）Aluminum alloy bell cranks with pressed in bushings, landing gear shock struts with pipe thread type grease fittings, clevis pin joints, shrink fits, and overstressed tubing B-nuts are examples of parts which are susceptible to stress corrosion cracking.

（5）In a predominately marine environment (with exposure to sea water and salt air), moisture-laden air is considerably more detrimental to an aircraft than it would be if all operations were conducted in a dry climate.

Lesson 8 Non-Destructive Testing

INTENSIVE READING

Non-Destructive Testing

Introduction

As aircraft increased in complexity, the rising cost of down time became a growing concern to aircraft operators. Airlines could not afford to ground an aircraft for long periods to conduct maintenance or inspections. Furthermore, as components became more expensive, new methods of inspection had to be developed to allow inspection without disassembly or destruction of the part. The inspection techniques developed are known as nondestructive testing (NDT). NDT methods are techniques used both in the production and in-service environments without damage or destruction of the item under investigation. There are a variety of nondestructive inspection techniques available to help determine the extent and degree of damage. Each has its own strengths and weaknesses, and more than one method may be needed to produce the exact damage assessment required. This chapter discusses the various methods of NDT including the fundamentals of visual, tap testing, dye penetration, magnetic particle, eddy current, ultrasonic, radiographic inspections, thermal imaging, laser shearography and holographic laser interferometry.

Visual Inspection

The most fundamental method of inspecting aircraft structures and components is through visual inspection. This method is irreplaceable in certain circumstances and limited in others. In any case, nothing can be inspected visually unless it is uncovered and made visible. Visual inspection can be a quite powerful and often underrated technique for detecting damage in aircraft structures. Even low-energy impacts may leave a slight marring, paint scrape, or faint surface blemish on a part. A slight wave or ripple on the surface may indicate an underlying delamination or disbond. Indications of impact damage are: dents cracks, cracked paint, missing paint, holes, depressions, eroded edges. A light spot or "whitish" area on a fiberglass part may indicate trapped air, a resin-lean area, or a delamination.

The basic tools required to conduct a visual inspection include a good light, a mirror, and some form of magnifying glass. The inspection must be performed in an area with good lighting. Use a flashlight in shaded areas. Use magnifying glass and mirrors as necessary. Flashlights are typically used to give spot-type illumination to the inspection area. When searching for surface cracks with a flashlight, direct the light beam at a 5 to 45 degree angle to the inspection surface, towards the face (Figure 8 – 1). Do not direct the light beam at such an angle that the reflected light beam shines directly into the eyes. Keep the eyes above the reflected light beam during the inspection. Determine the extent of any cracks found by directing the light beam at right angles to the crack and tracing its length. Use a 10 – power magnifying glass to confirm the existence of a suspected crack. The "flat angle" or "grazing light" method of visual inspection involves examining parts at a very flat angle, towards a light source, and looking for the subtle shadows caused by minor waves from damage or defects. If this is not adequate, use other NDT techniques, such as penetrant, magnetic particle, or eddy current to verify cracks.

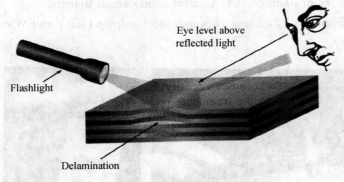

Figure 8 – 1   Visual inspection

A borescope is an optical device similar in principle to a telescope in that it enlarges objects like a magnifying glass. However, a borescope has a small lens mounted on a shaft with a built-in light source that illuminates the area being inspected. Borescopes are typically used to inspect inside engines using the spark plug hole for access.

Tap Testing

Tap test is also called a coin tap test. It can be performed using a metal disk, tap hammer, or an instrument. Tap testing is probably the most common inspection technique other than visual. Tap testing is widely used for a quick evaluation of any accessible aircraft surface to detect the presence of delamination or debonding.

The tap testing procedure consists of lightly tapping the surface of the part with a coin, light special hammer with a maximum of 2 ounces (see Figure 8 - 2), or any other suitable object. The acoustic response is compared with that of a known good area.

Lightly tap the area to be inspected, make a solid sound over areas that not delaminated or disbonded. Areas that are disbonded or delaminated will make a dull/hollow sound. A "flat" or "dead" response is considered unacceptable. The acoustic response of a good part can vary dramatically with changes in geometry, in which case a standard of some sort is required. The surface should be dry and free of oil, grease, and dirt. The entire area of interest must be tapped. By tapping back and forth over the area in question, and making a small mark at the point where the tone just begins to change, it is possible to outline large, irregularly-shaped areas of delaminations or disbonds. Tap testing is limited to finding relatively shallow defects in skins with a thickness less than .080 inch. In a honeycomb structure, for example, the far side bondline cannot be evaluated, requiring two-side access for a complete inspection. This method is portable, but no records are produced. The accuracy of this test depends on the inspector's subjective interpretation of the test response; therefore, only qualified personnel should perform this test. However, there are many limitations to tap testing, including: (1) Not good for deep damage. (2) Requires knowledge of the underlying structural detail of the part. (3) Not very effective in quantifying the degree of damage. (4) Cannot locate small defects.

As shown in Figure 8 - 2, using common tap hammer (left) and Wichitech RD3 digital tap hammer (right).

Figure 8 - 2　Tap testing

Dye Penetration

Dyepenetration inspection is a method of nondestructive inspection suitable for locating cracks, porosity, or other types of faults open to the surface in parts made of any nonporous material. Penetrant inspection is usable on ferrous and nonferrous metals, as well as nonporous plastic material. The primary limitation of dye penetrant inspection is that a defect must be open to the surface. This technique is only mentioned because it is quite effective and well-known as a non-destructive method of inspecting metal parts for tiny hairline cracks. However, it is not to be used for composites. Although this technique will also show where cracks have propagated in composites, the dye penetrant liquid contaminates the composite part. By definition, the dye liquid is designed to wick itself into tiny cracks, which will contaminate any bond-lines one might have in a repair scenario. Using this method eliminates any possibility of performing a bonded repair on the part. It can be useful on parts that will never be repaired, such as in a crash investigation, but it should not be used to assess the extent of damage in composite parts even though it is commonly used in metals.

Dye penetrant inspection is based on the principle of capillary attraction. The area being inspected is covered with a penetrating liquid that has a very low viscosity and low surface tension. This penetrant is allowed to remain on the surface long enough to allow the capillary action to draw the penetrant into any fault that extends to the surface. The smaller the defect, the longer the penetration time. After sufficient time, the excess penetrant is washed off and the surface is covered with a developer. The developer, by the process of reverse capillary action, blots the penetrant out of cracks or other faults forming a visible line in the developer. if an indication is fuzzy instead of sharp and clear, the probable cause is that the part was not thoroughly washed before the developer was applied (Figure 8 – 3).

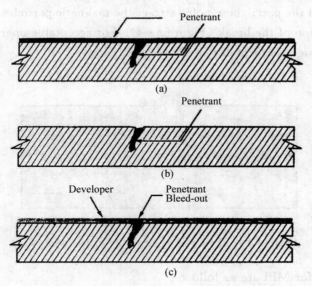

Figure 8 – 3   Dye penetration inspection

(a) When performing a liquid penetrant inspection, the penetrant is spread over the surface of the material being examined, and allowed sufficient time for capillary action to take place;

(b) The excess penetrant is then washed from the surface, leaving any cracks and surface flaws filled;

(c) An absorbent developer is sprayed over the surface where it blots out any penetrant. The crack then shows up as a bright line against the white developer

There are two types of dyes used in liquid penetrant inspection (LPI): fluorescent and colored. An ultraviolet light is used with the fluorescent penetrant and any flaw shows up as a green line. With the colored dye method, faults show up as red lines against the white developer.

Magnetic Particle Inspection

Magnetic particle inspection (MPI) is a method of detecting invisible cracks, laps, seams, voids, pits, subsurface holes, and other surface, or slightly subsurface, discontinuities in ferromagnetic materials, such as iron and steel. This method of inspection is a nondestructive test, which means it is performed on the actual part without damage to the part. It is not applicable to nonmagnetic materials. Surface and subsurface faults in a ferromagnetic part can be detected with magnetic particle inspection.

When a material containing large amounts of iron is subjected to a strong magnetic field, the magnetic domains within the material align themselves and the part becomes magnetized. When this happens, the part develops both a north anda south pole and lines of flux flow in a continuous stream from the north pole to the south pole. If a break occurs within the part another set of magnetic poles appears, one on either Side of the break. Therefore, when an oxide containing magnetic particles is poured or sprayed over the part's surface, any discontinuities in the material, either on or near the surface, create disruptions in the magnetic field around the part; these poles attract the magnetic particles in the oxide thereby giving you an indication of the break (Figure 8 – 4). For acceptable operation, the indicating medium must be of high permeability and low retentivity.

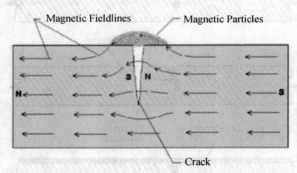

Figure 8 – 4　Principles of magnetic particle inspection

The basic steps for MPI are as follows:

1. Magnetize the part to be inspected. Any flaw or fault within the component interrupts the magnetic lines of flux and forms a north and a south pole.

2. Cover the area being inspected with very fine iron oxide particles.

3. The iron oxide is attracted to the magnetic poles where it forms a visible indication of the fault.

4. There are two ways of magnetizing a part. Overhaul manuals specify the way a part must be magnetized and the amount of current to be used for the magnetization. Circular

magnetization — by passing DC through the part; Lines of magnetic flux encircle the part at right angles to the flow of current and used for detecting faults that are parallel to the length of the part. Longitudinal magnetization — by holding the part inside a coil of wire carrying DC; Lines of flux extend lengthwise through the part at right angles to the coil and used for detecting faults that are perpendicular to the length of the part.

5. The iron oxide used to detect the fault contains a fluorescent dye. It may be applied as a dry powder, or as a suspension in a light oil such as kerosine.

6. The powder is dusted over the part, or the suspension is flowed over the surface being inspected. The oxide particles that are attracted to the poles created by the fault show up as a green mark when viewed under a black light.

7. Two types of magnetic particle inspection. (1) Continuous; the magnetizing current flows all the time the part is being inspected. (2) Residual; the part is magnetized and removed from the magnetic field, then inspected.

After inspection is completed, thoroughly demagnetize the part, in either of two ways; (1) Place the part in an AC magnetic field and slowly remove it from the field. (2) Place the part in a magnetic field made by pulses of DC of reversing polarity that is programmed to decrease its intensity. The reversing polarity of the field causes the magnetic domains within the material to continually change their orientation. The decreasing field strength allows them to remain in a disoriented condition.

Eddy Current Inspection

Eddy current is used to detect surface cracks, pits, subsurface cracks, corrosion on inner surfaces, and to determine alloy and heat-treat condition. Eddy current techniques are particularly well suited for detection of service-induced cracks in the field. Service-induced cracks in aircraft structures are generally caused by fatigue or stress corrosion. Both types of cracks initiate at the surface of a part. If this surface is accessible, a high-frequency eddy current inspection can be performed with a minimum of part preparation and a high degree of sensitivity. If the surface is less accessible, such as in a subsurface layer of structure, low-frequency eddy current inspection can usually be performed. Eddy current inspection can usually be performed without removing surface coatings such as primer, paint, and anodic films. Eddy current inspection has the greatest application for inspecting small localized areas where possible crack initiation is suspected rather than for scanning broad areas for randomly-oriented cracks. Pulsed eddy current testing technique received recognition and application in aviation industry. However, in some instances it is more economical to scan relatively large areas with eddy current rather than strip surface coatings, inspect by other methods, and then refinish. Eddy current inspection checks for faults inside a metal by detecting a change in its conductivity caused by the presence of a fault. This method is especially suited for detecting intergranular corrosion.

How it works? A test probe containing an AC excited coil induces an eddy current into the material being tested：(1) Excite the coil with the proper frequency of AC. (2) Place the probe on the surface being inspected so it can induce a changing magnetic field in the metal. (3) The changing magnetic field induces eddy currents in the metal. The amount of current is determined by four things：(1) The conductivity of the metal which is a function of its alloy type, grain size, degree of heat treatment, and tensile strength. (2) The permeability of the metal. (3) The mass of the material. (4) The presence of any faults or voids.

What it is suited for? (1) Identifying metals by comparison of their alloy type, degree of heat treatment, and tensile strength. (2) Detection of cracks or hidden faults. This is an ideal way to check aircraft wheels for cracks in the bead seat area. These cracks close up when the stress is off the wheel and are almost impossible to detect visually, but show up with eddy current inspection.

The methods of eddy current inspection are as follows：(1) Place the test probe on a piece of metal (known to be good) of the type being inspected, and zero the indicator. (2) Place the probe on the metal being inspected. If there are no internal faults, the indicator will again zero.

Eddy current inspection can be used of detection of corrosion. The mass of sound material changes when corrosion is present, either internally or on the opposite side of a skin being inspected. The steps are as follows：(1) Hold the eddy current probe against a part of the skin that is known to be free of corrosion and zero the meter. (2) Move the probe over the area being inspected. If corrosion is present, the meter will move off zero. (3) To inspect for corrosion around fastener holes, insert the small probe into a hole known to be free of corrosion and zero the indicator. When the probe is inserted into a hole where there is corrosion, the indicator will move off zero.

Ultrasonic Inspection

Ultrasonic inspection is used to detect flaws in a wide variety of materials, including metals and composites. It can be performed using portable battery-operated equipment, enabling parts to be inspected while still installed.

An ultrasonic testing (UT) instrument typically includes a pulser/receiver unit and a display device. The pulser/receiver unit includes a transducer probe that converts an electrical signal into a high frequency sound wave and then sends that wave into the structure being tested. A defect in the structure, such as a crack, causes a density change in the material and will reflect sound waves back to the transducer probe. The transducer converts the received sound waves (vibrations) into an electrical signal, which is then analyzed and shown on the UT display device. Inspection data may be displayed as an A-scan (time-based waveform display), B-scan (distance vs. time graph) or C-scan (plan view grayscale or color mapping), as shown in Figure 8 – 5.

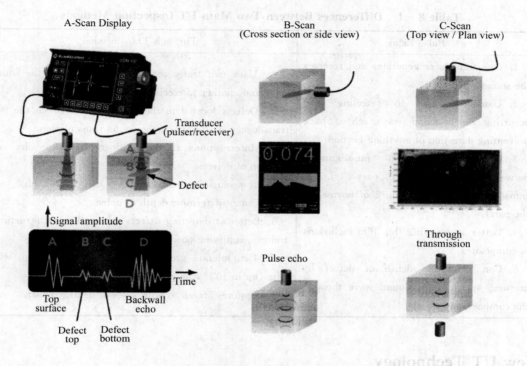

Figure 8 - 5   Ultrasonic testing

A-scans are the typical display provided, but C-scans are becoming more common. Interpretation of these displays requires extensive training, and a skilled, certified operator is mandatory. B-scans are not normally used in composites because they look from the side, in-plane with the plies, and thus do not reveal as much useful information as A-scans and C-scans. UT equipment will allow, under favorable circumstances, measurement of not only the location but also the depth of the damage, which cannot be done visually or by tap testing. However, there are many limitations, including:

1. Requires a couplant between the transducer probe and the part being inspected, as sound does not transmit well from air to solid materials. Couplant may be water or some type of gel.

2. Calibration standards are required for each type of material and thickness.

3. Interpretation of data demands careful analysis, requiring extensive skill, training and experience.

There are two basic techniques used in ultrasonic inspection: Pulse-echo and Through-transmission. Table 8 - 1 outlines the differences between these two methods.

Table 8 - 1   Differences Between Two Main UT Inspection Methods

Pulse-Echo	Through-Transmission
1. One transducer generates and receives the sound wave. 2. Usually limited to detecting first occurring defect-sound wave echoes back preventing detection of anything beyond. 3. More sensitive to misalignment between transducer and part (i. e. transducer should be within 2° of normal to the part). 4. Better at detecting thin film inclusions in composites. 5. Can determine depth of defect by knowing speed of the sound wave through the composite.	1. Uses two transducers: one to generate the sound wave and another to receive it. 2. Defects located in the sound path between the two transducers will interrupt the sound transmission. 3. Interruptions are analyzed to determine size and location of defects. 4. Less sensitive to small defects than pulse-echo. 5. Cannot determine depth of defect. 6. Better at detecting defects in multilayered structures and in quantifying porosity. 7. Can tolerate greater transducer misalignment with part (up to 10°F rom perpendicular). 8. Requires access to both sides of damaged part.

New UT Technology

Phased Array UT (Figure 8 - 6) is a relatively new offering, which may dramatically reduce the time required for inspection while providing excellent detection of small defects as well as the ability to determine both location and depth of defects. It achieves this by using an array of multiple transducers aligned in a single housing. By firing the transducers at slightly different times, the sound waves can be focused (depth) or steered (left, right, at an angle, etc.) toward a specific location. The transducer array is controlled electronically and can be programmed to sweep across and through the composite part. The transducer probe is also much larger than traditional UT probes, 4 "versus 0. 4", reducing the amount of time and movement across the surface required to provide 100% inspection of the part. The left is delaying elements in the center of the array changes the sound wave focus (depth), the right is delaying the timing of the left-side or right-side elements produces a sound wave inclined to the left or right, respectively.

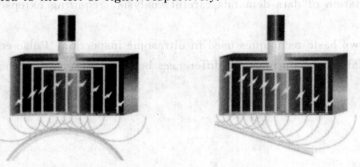

Figure 8 - 6   Phased Array UT Inspection

X-rays

X-rays are a form of radiation from beyond the ultraviolet end of the spectrum. They are generated by directing a stream of electrons at a metal target, usually tungsten. X-rays pass through the test specimen, and then are sensed on the other side by a detection device. Traditional detection devices included film-coated plates or a fluorescing screen. Current state-of-the-art is a computer-controlled system comprised of a flat-panel detector and digital image processing unit. X-rays are effective for locating hidden damage or presence of moisture, as shown in Figure 8 – 7.

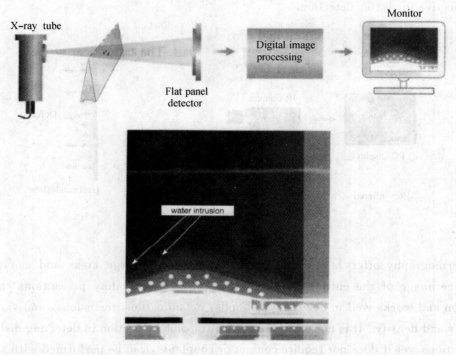

Figure 8 – 7　X-ray inspection display of wing flap from an F – 15 jet fighter, showing water trapped in honeycomb cells

X-ray inspection works better with fiberglass and aramid fiber composites than with carbon fiber. This is because carbon fiber and common matrix resins have very similar and very low X-ray absorption, resulting in an image that does not show much differentiation (i.e. appears black). X-ray inspection does work well in detecting transverse cracks, inclusions, honeycomb core damage and moisture ingression.

Because of the radiation involved, considerable safety protection is required when using this technique. Traditionally, the equipment was not very portable, but has become much more so over the past few years. The use of radio opaque penetrants helps to detect delamination and damage in complex structures, but also contaminates the part being inspected and eliminates the possibility of performing a bonded repair. In order to produce meaningful results, X-ray inspection requires a skilled, certified operator with significant training and experience.

Thermal Imaging

This technique uses a heat source (typically a flash gun or heat lamps) to heat the composite part being inspected. As the part cools down, it is monitored with an infrared camera and digital processing equipment. Irregularities in the temperature distribution across the surface indicate the presence of defects, as shown in figure 8 - 8. Infrared thermal imaging is an advanced NDT means. Infrared thermography inspection. (heat scanning) is more effective for water detection.

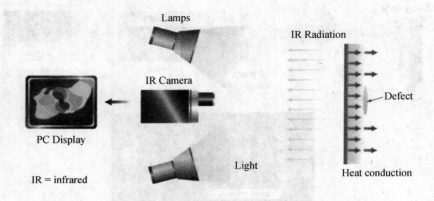

Figure 8 - 8　Thermal imaging

Thermography offers a quick method for inspecting large areas and can provide a subsurface image of the entire structure. It requires fewer safety precautions than X-ray inspection and works well in detecting disbonds, delamination, inclusions, and variations in thickness and density. It is not as sensitive as ultrasonic inspection in detecting disbonds and delaminations, yet it does not require contact or couplants, can be performed with single-side access (i. e. part installed), and is now achievable with portable equipment that is easy to set up and break down.

Three different inspection displays of the same panel having carbon fiber faceskins and Teflon inserts embedded in its aluminum honeycomb core, as shown in Figure 8 - 9.

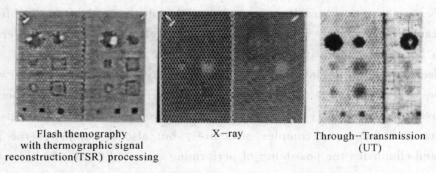

Figure 8 - 9　Thermal imaging vs. X-ray and UT inspection

Laser Shearography

Shearography nondestructive testing uses an image-shearing interferometer to detect and measure local out-of-plane deformation on the test part surface, as small as 3 nanometers, in response to a change in the applied engineered load. Shearography images tend to show only the local deformation on the target surface due to the presence of a surface defects subsurface flaws, delaminations, core damage, core splice joint separations, and impact damage.

Typical applied loads to the test part are dependant on the material reaction to the induced load. The optimum load type and magnitude depend on the flaw type depth, and is best determined before serial testing by making trial measurements. The applied load can be any of the following: heat, mechanical vibration, acoustic vibration, pressure and/or vacuum, electric fields, magnetic fields, microwave or mechanical load. Shearography offers exceptional high throughput and the ability to test a wide range of materials and structures including solid carbon fiber reinforced laminates, Composite Over-wrapped Pressure Vessels (COPVs), composite and metal honeycomb, foam cored sandwich structures and foam insulation of rocket launch vehicles.

Shearography is used both in production on new aerospace vehicles as well as during maintenance inspections. Applications include: AWACS rotodome, metal honeycomb control surfaces, radomes, air brakes, helicopter blades, aluminum and titanium honeycomb panels and structures, foam cored struc-tures and sandwich panels.

Portable shearography systems are designed for inspection of composite scarf joint repairs, boron repair patches, and non-visible impact damage on composite structures, particularly those with honeycomb core and thin facesheets. The entire system fits inside a transit case. The shearography inspection heads vacuum-attaches in any orientation to the aircraft surface. Stored inspection procedures can be called-up to test specific parts that have been pre-programmed by an operator or Level III engineer.

Holographic Laser Interferometry (HLI)

Holographic interferometry is a non-contact optical NDT technique that provides visual representation of the out-of-plane deformations of an object. The out-of-plane deformations appear as bright and dark fringes superimposed on a three-dimensional image of the object being interrogated. Object interrogation is performed by loading the object in some form (mechanically, thermally or surface pressure) and exposing a holographic plate to two half-exposures of laser light. The two exposures are done at different load states. See Figure 8 - 10.

Holographic interferometry is very sensitive to surface movement. The differential height between two bright fringe lines is half the laser light wavelength. Using a He-Ne laser the image on the right shows height relief between two fringes of the order of 0.3 micro-meters (0.000,3 mm or 0.007,6 inch). Thus very small defects can be identified with holographic interferometry, such as weak bonding behavior. Holographic interferometry also shows how the surface is reacting to the presence of the defect that might be hidden within the object. For example, a holographic interferogram will show the behavior of a skin surface in a sandwich structure with damage on the other skin.

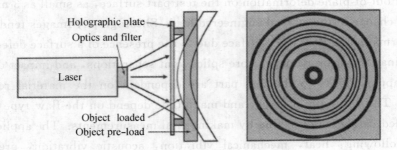

Figure 8 – 10　Concept of holographic laser interferometry

The holographic interferogram shown in Figure 8 – 11 was produced from initially placing a release film between the bondline of a thin sheet of aluminum and an aluminum plate. Note the fringe lines are uniform and further apart around the edges where the aluminum plates are well bonded. The center area depicts a disbonded area whereas the fringes are spaced closer together and in a less uniform manner. The partially bonded area exhibits fringes that are spaced randomly and indicate either a week bond or a partial disbond in this area.

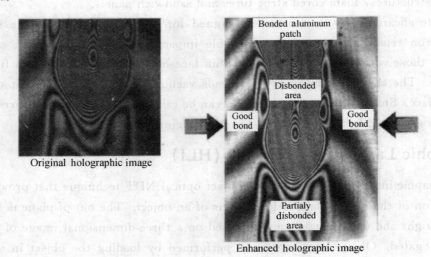

Figure 8 – 11　Actual HLI on aluminum plate

New Words/Phrases/Expression

1. nondestructive[ˌnɒndɪˈstrʌktɪv]　*adj.* 非破坏性的
2. visual[ˈvɪʒuəl]　*adj. & n.* 视觉的, 看得见的; 光学的画面, 图像

3. dye[daɪ] *n.* & *vt.* & *vi.* 染料,染色;颜色;染色;给……染色;染上或粘上(颜色)

4. penetration[ˌpenɪ'treɪʃən] *n.* 渗透;穿透;突破;洞察力

5. magnetic[mæg'netɪk] *adj.* 有磁性的,有吸引力的;磁性的;有吸引力的;有魅力的

6. eddy current['edi ˌkərənt] 涡流,涡电流

7. ultrasonic[ˌʌltrə'saːnɪk] *adj.* & *n.* 超声的;超声波的;超声速的;超声波

8. radiographic[ˌreɪdɪoʊ'græfɪk] *adj.* X射线照相术的

9. thermal imaging 热成像

10. laser shearography 激光剪切测量技术;激光错位散斑干涉测量技术

11. holographic[ˌhɑːlə'græfɪk] *adj.* 全息的

12. underrate[ˌʌndə'reɪt] *vt.* 对(某人或某事物)评价过低;看轻;轻视

13. mar['maːr] *vt.* & *n.* 毁坏;损坏;弄糟;糟蹋;玷污;污点;瑕疵;障碍;损伤;毁损

14. blemish['blemɪʃ] *n.* & *vt.* 瑕疵;缺点;污点;不名誉;玷污;损害;弄脏

15. delamination[diːˌlæmə'neɪʃən] *n.* 分层;分叶;层离;起鳞

16. flashlight['flæʃˌlaɪt] *n.* 手电筒;闪光信号灯

17. borescope['bɔːskəup] *n.* 管道镜

18. corrosion[kə'rəuʒən] *n.* 腐蚀;侵蚀;锈蚀;受腐蚀的部位;衰败

19. tap testing 敲击测验,敲击检查

20. debond[dɪ'bɒnd] *v.* 脱胶

21. acoustic[ə'kustɪk] *adj.* 听觉的;声学的;音响的

22. developer[dɪ'veləpər] *n.* 开发者;[摄]显影剂,显像剂

23. permeability[ˌpəmɪə'bɪlɪti] *n.* 渗透性;磁导率;可渗透性

24. retentivity[ˌriːten'tɪvəti] *n.* 保持力;[物]顽磁性

25. demagnetize[diː'mægnɪˌtaɪz] *vt.* 消磁,使退磁

26. transducer[trænz'djuːsər] *n.* 传感器,变频器,变换器

27. pulse-echo['pʌls'ekoʊ] *n.* 脉冲回波

28. spectrum['spektrəm] *n.* 光谱;波谱;范围;系列

29. radiation[ˌredɪ'eʃən] *n.* 辐射;放射物;辐射状;分散

30. thermography[θə'mɒɡrəfɪ] *n.* 热谱;温度记录;自记温度

31. portable['pɔːrtəbl] *adj.* & *n.* 手提的;轻便的;手提式打字机

32. interferometry[ˌɪntəfɪr'rɒmɪtrɪ] *n.* 干涉测量

Notes

(1) Indication of impact damage are: dents, cracks, cracked paint, missing paint, holes, depressions, eroded edges.

翻译:冲击损伤的迹象如下:凹坑、裂纹、漆层开裂、掉漆、穿孔、沉陷、边缘侵蚀。

(2) Visual inspection — The inspection must be performed in an area with good lighting. Use a flashlight in shaded areas. Use magnifying glass and mirrors as necessary.

翻译:目视检查——检查区域必须光线充足。在阴暗区域需使用手电筒,必要时需使用放

大镜和检查镜。

（3）Tap test is also called a coin tap test，it can be performed using a metal disk，tap hammer，or an instrument.

翻译：敲击测试又称为硬币敲击法，可以通过金属圆板、敲击锤或敲击仪器来实现。

（4）Lightly tap the area to be inspected，make a solid sound over areas that not delaminated or disbonded. Areas that are disbonded or delaminated will make a dull/hollow sound.

翻译：轻敲待检查区域，未分层/脱胶的区域会有坚实的声音，反之则会听到低沉、空洞的声音。

（5）Radiographic inspection(X-ray) — Effective for locating hidden damage or presence of moisture.

翻译：X射线能有效地探测到隐藏的损伤或水分。

（6）Infrared thermal imaging is an advanced NDT means. Infrared thermography inspection(heat scanning)is more effective for water detection.

翻译：红外热成像是一种先进的无损检测手段。红外扫描检查能更有效地检测水分的存在。

（7）Pulsed eddy current testing technique received recognition and application in aviation industry.

翻译：脉冲涡流检测技术在航空工业领域得到了广泛的重视和应用。

Exercises

Ⅰ. Direction：choose the right meaning for each item.

（1）显影时间
A. drain time B. developing time C. diffusion time

（2）着色渗透剂
A. dye penetrant B. dry developer C. drying oven

（3）涡流探伤仪
A. eddy current flaw detector
B. eddy current testing
C. imaging line scanner

（4）图像清晰度
A. image enhancement B. image magnification C. image definition

（5）胶片感光度
A. film processing B. film speed C. film contrast

（6）荧光渗透剂
A. fluorescent penetrant B. fluorescent light C. fluoroscopy

（7）全息照相
A. holography B. hydrophilic C. harmonics

（8）纵波探头

A. longitudinal wave probe　　B. longitudinal resolution　　C. longitudinal field

（9）磁场

A. magnetic flux　　　　　　B. magnetic force　　　　　　C. magnetic field

（10）显影过度

A. over development　　　　　B. over emulsfication　　　　C. over washing

（11）脉冲回波法

A. pulse echo method　　　　　B. pulse amplitude　　　　　C. pulse energy

（12）穿透法

A. through transmission technique

B. through penetration technique

C. through put

Ⅱ. Fill the right information in the form.

No.	Term	Abbreviation	Chinese
（1）	Nondestructive testing	NDT	
（2）	Liquid penetrant inspection	LPI	
（3）	Magnetic particle inspection	MPI	
（4）	Ultrasonic testing	UT	

Ⅲ. Choose the best answer in accordance with the text.

（1）The purpose of non-destructive inspection is：

A. To check for cracks in aircraft parts only.

B. To check for flaws and defects in a variety of different materials.

C. To check for flaws in non-ferrous metals only.

D. To check for flaws in ferrous metals only.

（2）Dye penetrant inspection may not be used to inspect ferrous metal parts for surface cracks.

A. True　　　　　　　　　　B. False

（3）A longitudinal crack in an engine wrist pin may be found with the magnetic particle inspection method by circularly magnetizing the pin.

A. True　　　　　　　　　　B. False

（4）What is recommended for positive identification of suspected cracks?

A. A mirror　　　　　　　　B. A 10×magnifying glass

C. A boroscope　　　　　　　D. A 2.5×magnifying glass

（5）Which method of magnetization would be used to detect a crack extending around a tubular ferrous metal part?

A. Circular magnetism　　　　B. Either method is equally effective

C. Longitudinal magnetism　　D. Either method as long as it is A. C.

（6）A fault located by the fluorescent penetrant method will show up as a（　　）.

— 143 —

A. Red   B. Green   C. White   D. Blue

(7) The material used to bring the penetrant back to the surface from a defect is called (   ).

A. reverse capillary action   B. fluorescents

C. emulsifier   D. developer

(8) Magnetic particle inspection is used primarily to detect (   ).

A. distortion   B. irregular surfaces

C. deep surface flaws   D. flaws on or near the surface

(9) Which non-destructive testing method is most successful in detecting intergranular corrosion in nonferrous metal?

A. Spectrographic inspection   B. Radiographic inspection

C. Magnetic particle inspection   D. Eddy current inspection

(10) What is the principal advantage of the radiographic inspection method of N. D. I. ?

A. Minimum safety precautions required.

B. Little or no disassembly of structure.

C. Simplicity of equipment operation.

D. Low cost.

Ⅳ. **Answer the following questions.**

(1) What inspection is the most fundamental method of inspecting aircraft structures?

(2) What tool is typically used to inspect inside engines using the spark plug hole for access?

(3) Which principle is dye penetrant inspection based on?

(4) There are two types of dyes used in liquid penetrant inspection, what are they?

(5) Why should a part be magnetized both longitudinally and circularly?

(6) If the flaw is less accessible, such as in a subsurface layer of structure, can low-frequency or high-frequency eddy current inspection usually be performed?

(7) Eddy current inspection cann't usually be performed without removing surface coatings such as primer, paint, and films, can it?

(8) Has ultrasonic detection equipment made it possible to locate defects in all types of materials without damaging the material being inspected? And about other types of NDT?

(9) There are two basic ultrasonic systems, what are they?

Ⅴ. **Translating the following sentences into Chinese.**

(1) As aircraft increased in complexity, the rising cost of down time became a growing concern to aircraft operators. Airlines could not afford to ground an aircraft for long periods to conduct maintenance or inspections.

(2) This chapter discusses the various methods of NDT including the fundamentals of visual, tap testing, dye penetration, magnetic particle, eddy current, ultrasonic, radiographic inspections, thermal imaging, laser shearography and holographic laser

interferometry.

(3) The most fundamental method of inspecting aircraft structures and components is through visual inspection. This method is irreplaceable in certain circumstances and limited in others.

(4) Tap testing is probably the most common inspection technique other than visual. Tap testing is widely used for a quick evaluation of any accessible aircraft surface to detect the presence of delamination or debonding.

(5) Dye penetration inspection is a method of nondestructive inspection suitable for locating cracks, porosity, or other types of faults open to the surface in parts made of any nonporous material. Penetrant inspection is usable on ferrous and nonferrous metals, as well as nonporous plastic material.

(6) Eddy current is used to detect surface cracks, pits, subsurface cracks, corrosion on inner surfaces, and to determine alloy and heat-treat condition. Eddy current techniques are particularly well suited for detection of service-induced cracks in the field. Service-induced cracks in aircraft structures are generally caused by fatigue or stress corrosion. Both types of cracks initiate at the surface of a part.

EXTENSIVE READING

Detection of Corrosion

Visual inspection

Exotic inspection equipment is often needed for certain parts of an aircraft to make the work of the A&P more efficient, but a well-trained eye is still the most effective tool for inspection. Corrosion can often be detected by careful visual inspection of the airplane structure. Corrosion of aluminum or magnesium appears as a white or gray powder along the edges of the skins and around rivet heads. Small blisters appearing under the finish on painted surfaces is an indication of corrosion. Examine the lap joints in the skins for bulges which indicate the presence of corrosion between the laying surfaces. Corrosion salts have more volume than sound aluminum and will push out against the skin.

The complex structure of modern aircraft makes the use of magnifying glasses, mirrors, bore-scopes, fiber optics, and other optical inspection tools imperative for a good visual inspection.

Dye penetrant inspection

Stress corrosion cracks are sometimes difficult, if not impossible, to detect by visual inspection alone, but they may be found by the use of a dye penetrant inspection. This inspection method is effective on both ferrous and non-ferrous metals and on non-porous plastics. The procedure used in dye penetrant inspection consists of spraying a penetrating liquid on the surface to be inspected. The liquid has a very low surface tension and will seep deep into any crack that extends to the surface. The penetrant is left on the surface long enough to penetrate all of the cracks and then the surface is wiped clean. A developer, which is a white, chalky powder, is sprayed over the surface where it acts as a blotter and draws the penetrant out of any cracks into which it has seeped.

The penetrant is usually dyed bright red, and so the cracks appear as red lines on the white surface. Another type of penetrant inspection uses a fluorescent penetrant which is viewed under an ultraviolet, or "black" light. This special light will cause any crack to appear as a green line on the surface. The main limitation of dye penetrant inspection is the fact that it can fail to find cracks that are so full of corrosion products that the dye cannot penetrate. Also, if the crack is filled with oil or grease, the penetrant cannot get in, and there will be no indication of a flaw. Porous or rough surfaces are almost impossible to clean of all of the penetrant, so materials with rough surfaces do not lend themselves to this type of inspection.

Ultrasonic inspection

One inspection method that is effectively used for corrosion inspection is that using ultrasonic energy. In this method of inspection, high-frequency pulses of energy, similar to sound waves, only at frequencies far above the audible range, are introduced into the airplane structure. There are two types of ultrasonic indications which may be used for corrosion detection: the pulse-echo and the resonance method.

In the pulse-echo method, a pulse of ultrasonic energy is directed into the structure by a device known as a transducer. This energy travels through the material to its opposite side and then bounces back. When the return pulse is received by the transducer, it is displayed on the screen of a cathode-ray oscilloscope as a spike which establishes a time base, representing the thickness of the material. If there is any change in thickness, such as may be caused by corrosion, the return will occupy a shorter space and will indicate the extent of damage. If there should be a crack or other flaw within the material, such as may be caused by intergranular corrosion, a second spike will appear on the oscilloscope screen which indicates the approximate position of the flaw within the material.

The second method of inspection using ultrasonic energy is the resonance method. This method operates on the principle that for any given thickness of material, there is a specific frequency of ultrasonic energy that will resonate, or produce the greatest amount of return. Variable frequency ultrasonic energy is fed into the transducer, and the output is monitored visually with a meter or audibly with a set of head phones. When the resonant frequency is reached, the meter will read the highest value or the tone will be the loudest in the phones. If the metal has been eaten away by corrosion, its resonant frequency will be different from that of sound metal, and the meter reading or tone volume will be lower.

The resonance method can be used to determine the actual thickness of the material by calibrating the probe with a test speciman of the same type material being tested. Ultrasonic inspection must be conducted by a person highly qualified and equipped for this procedure. False returns can easily disguise a fault, and many special transducers are needed for the different locations to be inspected.

Ultrasonic inspection is primarily a method of inspection by comparison. Good skin is measured, and any change in thickness of the skin being inspected may have been caused by corrosion. Further inspection must be made to determine the actual extent of the corrosion indicated by the ultrasonic inspection.

Radiological inspection

Like ultrasonic inspection, radiological inspection such as X-ray is used to determine, from the outside of the structure, if there is any damage on the inside. X-ray is a photographic form of inspection in which extremely high-frequency pulses of electromagnetic radiation are passed through the structure being inspected. Energy at this frequency has the ability to expose photographic film, and as it passes through the structure, areas of high density pass less of the radiation energy and therefore expose the film less. After the exposure, the film is developed like any other photographic negative. In areas where the density of the material is greatest, less energy will penetrate, and this area will appear light on the negative. Areas of lower density pass more energy, expose the film more, and appear darker. X-ray inspection requires extensive training and experience for proper interpretation of the results. The use of X-ray involves some danger, because exposure to the radiation energy used in this process may cause burns, damage to the blood, and possibly death. Persons around X-ray equipment should wear a radiation monitor film badge which is developed at the end of an exposure time to determine the amount of radiation the wearer has absorbed. A blood count should be made periodically for persons involved in X-ray inspections.

The depth of penetration of X-rays is determined by the amount of power used. This power is normally measured in terms of kilovolts applied to the X-ray tube and may range from about eight kilovolts to as high as 200 kilovolts. The lower power applications, called "soft" X-rays, are used to inspect for corrosion.

Comparison of NDT Techniques

Table 8 - 2 provides a list of the advantages and disadvantages of common NDT methods. Table 8 - 2, in conjunction with other information in the AC, may be used as a guide for evaluating the most appropriate NDT method when the manufacturer or the FAA has not specified a particular NDT method to be used.

Table 8 - 2 Advantages and disadvantages of NDT methods

Method	Advantages	Disadvantages
Visual	Inexpensive Highly portable Immediate results Minimum training Minimum part preparation	Surface discontinuities only Generally only large discontinuities Misinterpretation of scratches
Dye Penetrant	Portable Inexpensive Sensitive to very small discontinuities 30 min. or less to accomplish Minimum skill required	Locate surface defects only Rough or porous surfaces interfere with test Part preparation required (removal of finishes and sealant, etc.) High degree of cleanliness required Direct visual detection of results required
Magnetic Particle	Can be portable Inexpensive Sensitive to small discontinuities Immediate results Moderate skill required Detects surface and subsurface discontinuities Relatively fast	Surface must be accessible Rough surfaces interfere with test Part preparation required (removal of finishes and sealant, etc.) Semi-directional requiring general orientation of field to discontinuity Ferro-magnetic materials only Part must be demagnetized after test.
Eddy Current	Portable Detects surface and subsurface discontinuities Moderate speed Immediate results Sensitive to small discontinuities Thickness sensitive Can detect many variables	Surface must be accessible to probe Rough surfaces interfere with test Electrically conductive materials Skill and training required Time consuming for large areas
Ultrasonic	Portable Inexpensive Sensitive to very small discontinuities Immediate results Little part preparation Wide range of materials and thickness can be inspected	Surface must be accessible to probe Rough surfaces interfere with test Highly sensitive to sound beam — discontinuity orientation High degree of skill required to set up and interpret Couplant usually required
X-ray Radiography	Detects surface and internal flaws Can inspect hidden areas Permanent test record obtained Minimum part preparation	Safety hazard Very expensive (slow process) Highly directional, sensitive to flaw orientation High degree of skill and experience required for exposure and interpretation Depth of discontinuity not indicated

Continue

Method	Advantages	Disadvantages
Isotope Radiography	Portable Less expensive than X-ray Detects surface and internal flaws Can inspect hidden areas Permanent test record obtained Minimum part preparation	Safety hazard Must conform to Federal and State regulations for handling and use Highly directional, sensitive to flaw orientation High degree of skill and experience required for exposure and interpretation Depth of discontinuity not indicated

As shown in Table 8 – 3, each NDT method has its own strengths and weaknesses, and more than one method may be needed to produce the exact damage assessment required.

Table 8 – 3   Damage Inspection Methods

	Visual	Tap Test	A-Scan	C-Scan	X-Rays	Thermal imaging	Holographic laser interferometry	Laser shearography
Surface Deiaminations	◖	●	◖	●	◖	●	●	◖
Deep Delaminations	N/A	○	●	●	●	◖	◖	◖
Full Disbond	◖	◖	●	●	◖	●	●	◖
Kissing Disbond	N/A	○	○	○	N/A	N/A	●	◖
Core Damage	◖	◖	○	●	●	◖	●	◖
Inclusions	◖	◖	●	●	●	●	◖	◖
Porosity	◖	N/A	◖	●	N/A	N/A	◖	N/A
Voids	◖	●	◖	◖	◖	◖	●	◖
Backing Film	N/A	◖	◖	◖	◖	◖	◖	N/A
Bdge Damage	●	◖	◖	●	●	◖	●	●
Heat Damage	◖	◖	●	◖	N/A	◖	●	N/A
Severe Impact	●	●	●	●	○	●	●	●
Medlum Impact	●	●	●	●	N/A	○	●	●
Minor Impact	○	○	○	○	N/A	○	●	●
Uneven Bondline	○	N/A	○	○	○	○	◖	N/A
Weak Bond	N/A	N/A	N/A	N/A	N/A	N/A	◖	N/A
Water in Core	N/A	◖	○	●	◖	●	N/A	N/A

● Excellent   ◖ Good   ○ Poor

Lesson 9 Aircraft Fasteners

INTENSIVE READING

Common Aircraft Fasteners

In addition to the numerous permanent fasteners used on aircraft, there is a second type of fastener that, unlike rivets, can be reused. These fasteners include threaded hardware such as bolts and screws and the various types of nuts that secure them.

Thread Fasteners

Threaded fasteners allow parts to be fastened together with all of the strength unthreaded fasteners provide. However, unlike rivets and pins, threaded fasteners may be disassembled and reassembled an almost infinite number of times. In addition to being identified as either coarse or fine, threads are also designated by class of fit from one to five. Aircraft bolts are usually fine threaded with a Class 3 fit (medium fit), whereas screws are typically a Class 2 (free fit) or 3 fit.

Bolts are identified by their diameter and length. A diameter represents the shank diameter while the length represents the distance from the bottom of the head to the end of the bolt. A bolt's grip length is the length of the unthreaded portion(Figure 9 – 1).

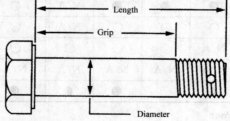

Figure 9 – 1   When choosing an AN bolt for a specific application, you must know the
diameter, length, and grip length required

In addition to the designation code, most aircraft bolts have a marking on their head identifying what the bolt is made of and, in many cases, the manufacturer. For example, a standard AN bolt has an asterisk in the center of its manufactured head. A raised dash means corrosion-resistant steel and two raised dashes means 2024 alloy aluminum (Figure 9 – 2).

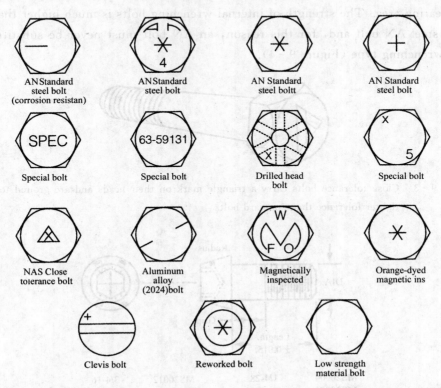

Figure 9 – 2   To aid in identifying what type of metal a bolt is made of, bolt heads are marked with a symbol

Drilled-Head Engine Bolts

Drilled-head engine bolts have a thicker head that is drilled with a small hole in each of the flats and in the center of the head. An advantage of drilled-head engine bolts is that they are made with either fine or coarse threads.

Close Tolerance Bolts

Close tolerance bolts are ground to a tolerance of + 0.000 – 0.000, 5 inch. Close tolerance bolts must be used in areas that are subject to pounding loads or in a structure that is required to be both riveted and bolted (Figure 9 – 3).

Internal Wrenching Bolts

MS20004 through MS20024 internal wrenching bolts are high-strength steel bolts used primarily in areas that are subjected to high tensile loads. A six-sided hole is machined into the center of their heads to accept an Allen wrench of the proper size. These bolts have a radius between the head and shank and, when installed in steel parts, the hole must be counterbored to accommodate this radius. When an internal wrenching bolt is installed in an aluminum alloy structure, a MS20002C washer must be used under the head to provide the needed bearing area. The strength of internal wrenching bolts is much higher than that of a standard steel AN bolt and, for this reason, an AN bolt must never be substituted for an internal wrenching type (Figure 9 - 4).

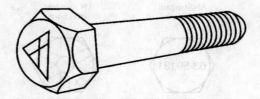

Figure 9 - 3   Close tolerance bolts carry a triangle mark on their heads and are ground to a much tighter tolerance than standard bolts

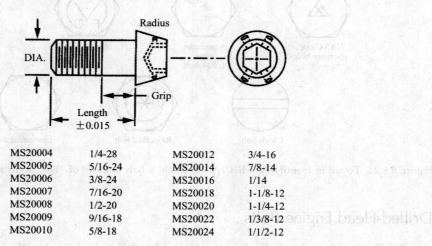

MS20004	1/4-28	MS20012	3/4-16
MS20005	5/16-24	MS20014	7/8-14
MS20006	3/8-24	MS20016	1/14
MS20007	7/16-20	MS20018	1-1/8-12
MS20008	1/2-20	MS20020	1-1/4-12
MS20009	9/16-18	MS20022	1/3/8-12
MS20010	5/8-18	MS20024	1/1/2-12

Figure 9 - 4   High-strength internal wrenching bolts can bear high tension loads and are frequently used to mount engines

Bolt Selection and Installation

When joining two pieces of material, their combined thickness determines the correct length of bolt to use. If the grip length is slightly longer than this thickness, washers must be added to ensure that the nut can provide the proper amount of pressure when it is tightened. On the other hand, if the grip length is substantially less than the thickness of the materials the bolt's threads will extend into the material, resulting in a weaker joint. Unless otherwise specified in an assembly drawing, bolts should be installed with their head on top or forward. Placing the head in either of these positions makes it less likely that a bolt will fall out of a hole if the nut is lost. An acronym to help remember the proper direction for bolt installation is "IDA", which stands for inboard, down, or aft.

Nuts

All nuts used in aircraft construction must have some sort of locking device to prevent them from loosening and falling off. Many nuts are held on a bolt by passing a cotter pin through a hole in the bolt shank and through slots, or castellations, in the nut. Others have some form of locking insert that grips a bolt's threads or relies on the tension of a spring-type lock-washer to hold the nut tight enough against the threads to keep it from vibrating loose. There are two basic types of nuts, self-locking and non self-locking. As the name implies, a self-locking nut locks onto a bolt on its own while a non self-locking nut relies on either a cotter pin, check nut, or lock washer to hold it in place (Figure 9 – 5).

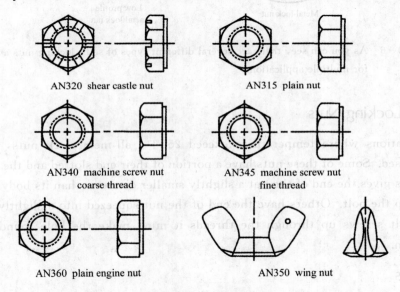

AN320 shear castle nut   AN315 plain nut

AN340 machine screw nut   AN345 machine screw nut
coarse thread   fine thread

AN360 plain engine nut   AN350 wing nut

Figure 9 – 5   Standard aircraft nuts are available for a variety of applications

Self-Locking Nuts

The two general types of self-locking nuts used in aviation are the fiber, or nylon type, and the all metal type (Figure 9 – 6).

Low-Temperature Self-Locking Nuts

AN365 self-locking nuts are used on bolts and machine screws and are held in position by a nylon insert above the threads. This insert has a hole slightly smaller than the thread diameter on which it fits. This creates friction between the threads and nut to keep the nut from vibrating loose. Nylon self-locking nuts should not be used in any location where the temperature could exceed 250°F.

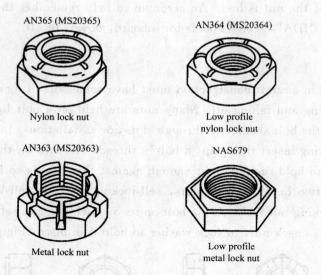

Figure 9 – 6   As you can see, there are several different types of self – locking nuts available for multiple applications

Metal Self-Locking Nuts

In applications where temperatures exceed 250°F, all-metal lock nuts, such as the AN363, are used. Some of these nuts have a portion of their end slotted and the slots swaged together. This gives the end of the nut a slightly smaller diameter than its body allowing the threads to grip the bolt. Others have the end of the nut squeezed into a slightly oval shape, and as the bolt screws up through the threads it must make the hole round, creating a gripping action.

Anchor Nuts

Anchor nuts are permanently mounted nut plates that enable inspection plates and access doors to be easily removed and installed (Figure 9 – 7).

Screws

Screws are probably the most commonly used threaded fastener in aircraft. They differ from bolts in that they are generally made of lower strength materials. Screws are typically installed with a loose-fitting thread, and the head shapes are made to engage a screwdriver or wrench. There are three basic classifications of screws used in aircraft construction: machine screws, which are the most widely used; structural screws, which have the same strength as bolts; and self-tapping screws, which are typically used to join light weight materials.

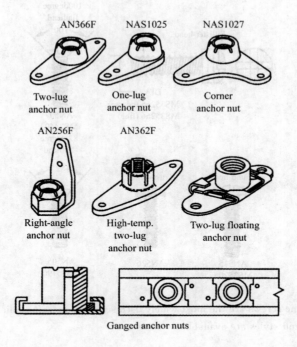

Figure 9 – 7   Anchor nuts simplify the process of installing and removing inspection plates. Some
of the more familiar anchor nuts are shown above

Machine Screws

Machine screws are used extensively for attaching fairings, inspection plates, fluid line
clamps and other light structural parts. The main difference between aircraft bolts and
machine screws is that the threads of a machine screw usually run the full length of the
shank, whereas bolts have an unthreaded grip length. Screws normally have a Class 2 or free
fit and are available in both national coarse and national fine threads. The most common
machine screws used in aviation are the fillister-head screw, the flat-head screw, the round-
head screw, and the truss-head screw (Figure 9 – 8).

Structural Screws

Structural screws are made of alloy steel, are heat treated, and can be used as structural
bolts. They have a definite grip and the same shear strength as a bolt of the same size.
Shank tolerances are similar to AN hex-head bolts, and the threads are National Fine.
Structural screws are available with fillister, flat, or washer heads.

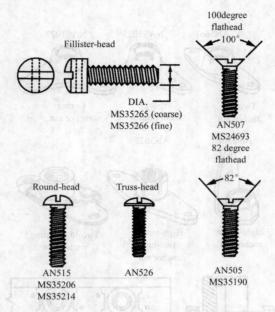

Figure 9 – 8　Machine screws are the most widely used screw in aircraft applications. A variety of sizes and styles are available

Self-tapping Screws

Self-tapping screws have coarse-threads and are used to hold thin sheets of metal, plastic, or ply-wood together. The type-A screw has a gimlet (sharp) point, and the type B has a blunt point with threads that are slightly finer than those of a type-A screw. There are four types of heads available on self-tapping screws: a round head, a truss head, a countersunk head, which is flat on top, and the countersunk oval screw. The truss-head is rounded, similar to the round head screw, but is considerably thinner (Figure 9 – 9).

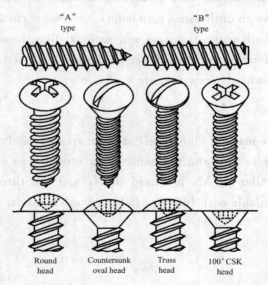

Figure 9 – 9　Self-tapping sheet metal screws are useful for attaching trim and upholstery

Pins

The main types of pins used in aircraft structures are the roll pin, clevis pin, cotter pin, and taper pin. Pins are used in shear applications and for safety (Figure 9 – 10 – Figure 9 – 13).

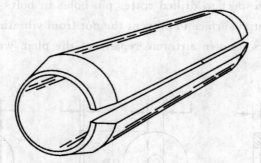

Figure 9 – 10   MS16562 spring steel roll pins are often used in the movable joints of aircraft seats

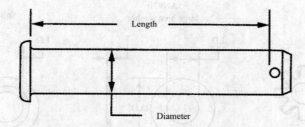

Figure 9 – 11   AN392 through AN406 (MS20392) clevis pin are often found in control cable systems

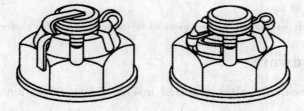

Figure 9 – 12   You may use either of the two acceptable methods to safety castellated nuts

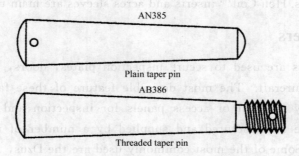

Figure 9 – 13   Taper pins produce a tight fit in a reamed hole for applications loaded in shear

Washers

Washers provide a bearing surface area for nuts, and act as spacers or shims to obtain the proper grip length for a bolt and nut assembly. They are also used to adjust the position of castellated nuts with respect to drilled cotter pin holes in bolts as well as apply tension between a nut and a material surface to prevent the nut from vibrating loose. The three most common types of washers used in airframe repair are the plain washer, lock washer, and special washer (Figure 9 – 14).

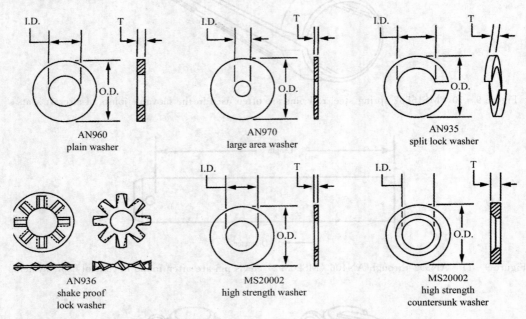

Figure 9 – 14   Aircraft washer come in a variety of styles and are used to perform a variety of tasks

Hole Repair Hardware

Threaded holes wear out after repeated insertion and extraction of fasteners. In the past, this often meant an expensive part had to be scrapped when a hole was stripped out. However, now, hole repair hardware allows you to make a fast and inexpensive repair to worn or damaged holes. Heli-Coil™ inserts and acres sleeves are main types.

Turnlock Fasteners

Turnlock fasteners are used to secure inspection plates, doors, cowlings, and other removable panels on aircraft. The most desirable feature of these fasteners is that they permit quick and easy removal of access panels for inspection and servicing purposes. Turnlock fasteners are manufactured and supplied by a number of manufacturers under various trade names. Some of the most commonly used are the Dzus, Airloc, and Camlock (Figure 9 – 15 – Figure 9 – 17).

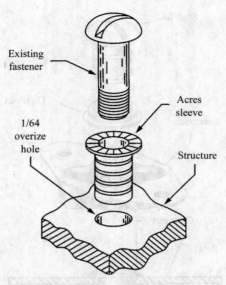

Figure 9 – 15   Acres sleeves are used to repair holes that have been worn oversize or corroded. The hole is drilled oversize and the sleeve is pressed or bonded in place. The original fastener is then installed

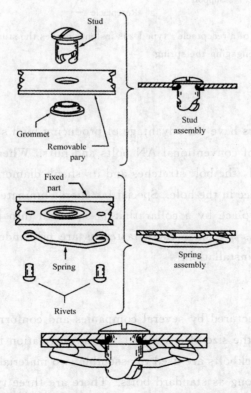

Figure 9 – 16   With a standard Dzus fastener, a slotted stud engages a spring mounted to the fuselage. As the stud is turn one quarter turn, the spring locks the fastener in place

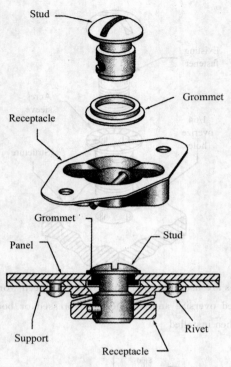

Figure 9 – 17   The receptacle of a receptacle – type Dzus fastener guides the stud to the exact location it needs to be prior to engaging the spring

Special Fasteners

Many special fasteners have the advantage of producing high strength with light weight and can be used in place of conventional AN bolts and nuts. When a standard AN nut and bolt assembly is tightened, the bolt stretches and its shank diameter decreases, causing the bolt to increase its clearance in the hole. Special fasteners eliminate this change in dimension because they are held in place by a collar that is squeezed into position instead of being screwed on like a nut. As a result, these fasteners are not under the same tensile loads imposed on a bolt during installation.

Lockbolts

Lockbolts are manufactured by several companies and conform to Military Standards. These standards describe the size of a lockbolt's head in relation to its shank diameter, as well as the alloy used. Lockbolts are used to assemble two materials permanently. They are lightweight and are as strong as standard bolts. There are three types of lockbolts used in aviation, they are the pull-type lockbolt, the blind-type lockbolt, and the stump-type lockbolt. Lockbolts are available for both shear and tension applications. With shear lockbolts, the head is kept thin and there are only two grooves provided for the locking collar. However, with tension lockbolts, the head is thicker and four or five grooves are provided to allow for higher tension values. The locking collars used on both shear and tension lockbolts are color coded for easy identification (Figure 9 – 18).

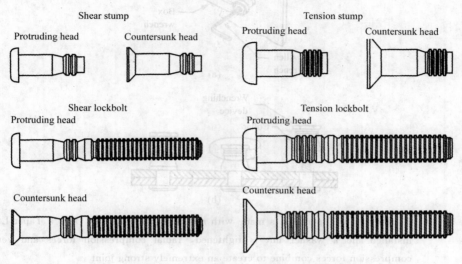

Figure 9 – 18   As you can see, both the shear and tension type lockbolts come in a variety of sizes with multiple head styles

Hi-Loks

Hi-Lok bolts are manufactured in several different alloys such as titanium, stainless steel, steel, and aluminum. They possess sufficient strength to withstand bearing and shearing loads, and are available with flat and countersunk heads. A conventional Hi-Lok has a straight shank with standard threads. Although wrenching lock nuts are usually used, the threads are compatible with standard AN bolts and nuts. To install a Hi-Lok, the hole is first drilled with an interference fit. The Hi-Lok is then tapped into the hole and a shear collar is installed. A Hi-Lok retaining collar is installed using either specially prepared tools or a simple Allen and box end wrench. Once the collar is tightened to the appropriate torque value, the wrenching device shears off leaving only the locking collar (Figure 9 – 19).

Hi-Lites

The Hi-Lite fastener is similar to the Hi-Lok except that it is made from lighter materials and has a shorter transition from the threaded section to the shank. Furthermore, the elimination of material between the threads and shank give an additional weight saving with no loss of strength. The Hi-Lite's main advantage is its excellent strength to weight ratio.

Hi-Lites are available in an assortment of diameters ranging from 3/16 to 3/8 inch. They are installed either with a Hi-Lok locking collar or by a swaged collar like the Lockbolt. In either case, the shank diameter is not reduced by stretch torquing.

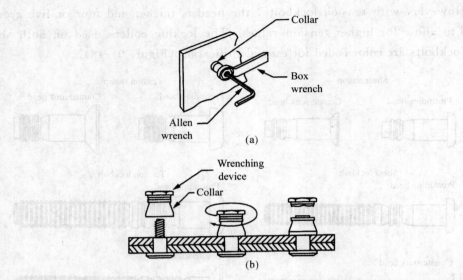

Figure 9 – 19　The hole of a Taper-Lok is made with a special tapered drill. Once a Taper-Lok is installed and a washer nut is tightened, radial compression forces and vertical compression forces combine to create an extremely strong joint

CherryBUCKs

The CherryBUCK is a one-piece special fastener that combines two titanium alloys which are bonded together to form a strong structural fastener. The head and upper part of the shank of a CherryBUCK is composed of 6AL – 4V alloy while Ti-Cb alloy is used in the lower shank. When driven, the lower part of the shank forms a bucktail.

An important advantage of the CherryBUCK is the fact that it is a one-piece fastener. Since there is only one piece, CherryBUCKs can safely be installed in jet engine intakes with no danger of foreign object damage. This type of damage often occurs when multiple piece fasteners lose their retaining collars and are ingested into a compressor inlet (Figure 9 – 20).

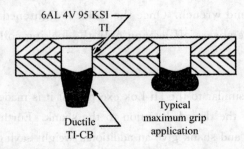

Figure 9 – 20　CherryBUCK fasteners combine two titanium alloys to produce a one-piece fastener with 95 KSI shear strength

Taper-Lok

The strongest special fastener, used in the construction of many standard and wide-body jet aircraft, is the Taper-Lok, a breakthrough in jet age fasteners. The Taper-Lok was developed to improve the structural fatigue life of modern jet aircraft. Because of its tapered shape, the Taper-Lok exerts a force on the conical walls of a hole, much like a cork in a wine bottle. To a certain extent, a Taper-Lok mimics the action of a driven solid shank rivet, in that it completely fills the hole. However, a Taper-Lok does this without the shank swelling.

When a washer nut draws the Taper-Lok into its hole, the fastener pushes outward and creates a tremendous force against the tapered walls of the hole. This creates radial compression around the shank and vertical compression lines as the metals are squeezed together. The combination of these forces generate strength unequaled by any other type of fastener (Figure 9 – 21).

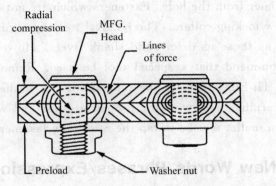

Figure 9 – 21   The hole for a Taper-Lok is made with a special tapered drill. Once a Taper-Lok is installed and a washer nut is tightened, radial compression forces and vertical compression forces combine to create an extremely strong joint

Hi-Tigue

The Hi-Tigue fastener has a bead that encircles the bottom of its shank and is a further advancement in special fastener design. This bead preloads the hole it fills, resulting in increased joint strength. During installation, the bead presses against the side wall of the hole, exerting a radial force which strengthens the surrounding area. Since it is preloaded, the joint is not subjected to the constant cyclic action that normally causes a joint to become cold worked and eventually fail. Hi-Tigue fasteners are produced in aluminum, titanium, and stainless steel alloys. The collars are also composed of compatible metal alloys and are available in two types, sealing and non-sealing. As with Hi-Loks, Hi-Tigues can be installed using an Allen and box end wrench (Figure 9 – 22).

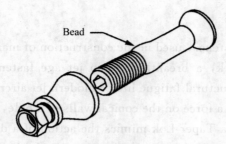

Bead

Figure 9 – 22   This Hi-Tigue features a subtly shaped bead at the threaded end of the shank. This bead preloads the hole it is inserted into thereby strengthening the joint

Removal of Special Fasteners

Special fasteners that are locked into place with a crushable collar are easily removed by splitting the collar with a small cape chisel. After the collar is split, knock away the two halves and tap the fastener from the hole. Fasteners which are not damaged during removal can be reused using new locking collars. The removal techniques of certain special fasteners are basically the same as those used for solid shank rivets. However, in some cases, the manufacturer may recommend that a special tool be used. Removal of Taper-Loks, Hi-Loks, Hi-Tigues, and Hi-Lites requires the removal of the washer nut or locking collar. Both are removed by turning them with the proper size box end wrench or a pair of vise-grips. After removal, a mallet is used to tap the remaining fastener out of its hole.

New Words/Phrases/Expression

1. fastener['fɑːsnə]   *n.* 紧固件
2. thread[θred]   *n.* 螺纹
3. bolt[bəult]   *n.* 螺栓
4. screw[skruː]   *n.* 螺钉、螺丝
5. nut[nʌt]   *n.* 螺母
6. pin[pin]   *n.* 销钉、销子
7. disassemble[,disə'sembl]   *v.* 分解
8. diameter[dai'æmitə]   *n.* 直径
9. asterisk['æstərisk]   *n.* 星号
10. tolerance['tɔlərəns]   *n.* 公差
11. close tolerance bolt   *n.* 高精度螺栓
12. triangle['traiæŋgl]   *n.* 三角形
13. Internal wrenching bolt   内六方螺栓
14. subject to   遭受
15. accomodate[ə'kɒmədeit]   *n.* 调节,调和

16. washer['wɔʃə]   *n.* 垫片

17. grip length   光杆长度,夹紧长度

18. substantially[səb'stænʃəli]   *adv.* 本质上,实质上

19. acronym['ækrənim]   *n.* 首字母缩略词

20. cotter pin   *n.* 开口销

21. castellation[kæstə'leiʃ(ə)n]*n.* 蝶形

22. vibrate['vaibreit]   *v.* 振动

23. self-locking   自锁

24. check nut   防松螺母

25. squeeze[skwi:z]   *v.* 挤压

26. lock washer   止动垫圈

27. oval['əuvəl]   *adj.* 椭圆形的

28. grip[grip]   *v.* 夹紧

29. anchor nut['æŋkə]   *n.* 托板螺母

30. loosing fitting thread   松配合螺纹

31. engage[in'geidʒ]   *v.* 啮合,配合

32. machine screw   机用螺钉,非结构螺钉

33. structure screw   结构螺钉

34. self-tapping screw   自攻螺钉

35. free fit   自由配合

36. national coarse thread   国家标准粗牙螺纹

37. national fine thread   国家标准细牙螺纹

38. fillister head screw   有槽圆头螺钉

39. flat head screw   平头螺钉

40. round head screw   半圆头螺钉

41. truss head screw   大圆头螺钉

42. AN(Army navy)   陆军海军

43. Hex-head bolt   外六角头螺栓

44. ply-wood   多层板

45. gimlet point   尖头

46. blunt point   平头

47. roll pin   柱形插销;滚销

48. movable joint   活接头

49. clevis pin   销钉

50. cotter pin   开口销

51. taper pin   锥形销;圆柱销

52. reamed hole   铰制孔

53. tight fit   紧密配合

54. plain washer   普通垫圈

55. lock washer   防松垫圈

56. special washer   专用垫圈,特殊垫圈

57. insert[in'sɜːt]   *n.* 嵌块

58. sleeve[sliːv]   *n.* 嵌套管,轴套

59. hole repair hardware   孔修补五金件

60. turnlock fastener   转锁紧固件

61. lockbolt   自锁螺栓

62. shear lockbolt   抗剪切自锁螺栓

63. tensile lockbolt   抗拉伸自锁螺栓

64. interference fit   干涉配合

65. retaining collar   止动环

66. transition[træn'ziʃ(ə)n]   *n.* 嵌过渡

67. breakthrough['breikθruː]   *n.* 嵌突破

68. preload[ˌpriː'ləud]   *n.* 嵌预加载

69. cyclic action   反复作用

70. assortment[ə'sɔːtmənt]   *n.* 嵌分类

Notes

(1)In addition to being identified as either coarse or fine, threads are also designated by class of fit from one to five. Aircraft bolts are usually fine threaded with a Class 3 fit (medium fit), whereas screws are typically a Class 2 (free fit) or 3 fit.

分析:In addition to 和 beside 同义,都可翻译成除……之外,但和 except 有所不同。

翻译:除了可将螺纹分为粗螺纹和细螺纹以外,螺纹还可按配合等级分为 1～5 级。航空螺栓通常是 3 级配合(中度配合)的细螺纹,而螺钉通常是 2 级配合(自由配合)或 3 级配合。

(2)As the name implies, a self-locking nut locks onto a bolt on its own while a non self-locking nut relies on either a cotter pin, check nut, or lock washer to hold it in place.

分析:此句的"while"之后应翻译出转折的意思。

翻译:就像名字里隐含的,自锁螺母通过自身锁紧在螺栓上,而非自锁螺母则需靠开口销、防松螺母或锁紧垫片使它上紧在紧固件上。

(3)Others have the end of the nut squeezed into a slightly oval shape, and as the bolt screws up through the threads it must make the hole round, creating a gripping action.

分析:此句为 have sth. done 的句型,此句中的 screws up 应翻译为旋合。

翻译:其他的(全金属式自锁螺母)是将其螺母尾部挤压成轻微的椭圆形,当螺栓旋合入(螺母的)螺纹,它必须使螺母的孔撑圆,这样就形成了夹紧作用。

(4)Machine screws are used extensively for attaching fairings, inspection plates, fluid line clamps and other light structural parts.

分析:被动语态。

翻译:机用螺钉(非结构螺钉)被广泛应用于安装整流罩,检查盖板,流体管线管夹和其他

较轻的结构零件。

（5）When a standard AN nut and bolt assembly is tightened，the bolt stretches and its shank diameter decreases，causing the bolt to increase its clearance in the hole.

分析：条件状语从句。

翻译：当一组标准陆军海军螺母和螺栓被拧紧时，螺栓伸长，直径减小，这样就使得螺栓与孔之间的缝隙增大。

Exercises

Ⅰ. Answer the following questions.

（1）What is the difference between thread fasteners and rivet?

（2）Which fit are aircraft bolts usually fine threaded with?

（3）How to identified bolts?

（4）Where are MS20024 used?

（5）How to select bolts when joining two pieces of material?

（6）In what condition are all-metal lock nuts used?

（7）What is the difference between screw and bolts?

（8）Which are the most widely used in the three basic classifications of screws?

（9）What are the common machine screws used in aviation?

（10）What's the function of pins and washers?

（11）What are turnlock fasteners used to do?

（12）What is the advantage of special fasteners?

（13）How many types of lockbolts are used in aviation?

（14）How to install a Hi-Lok?

（15）What is the important advantage of the CherryBUCK ?

（16）Which is the strongest special fastener?

（17）Name three conventional, non-blind, special fasteners.

Ⅱ. Put the following sentences into Chinese.

（1）Screws are typically installed with a loose-fitting thread，and the head shapes are made to engage a screwdriver or wrench.

（2）Structural screws are made of alloy steel，are heat treated，and can be used as structural bolts. They have a definite grip and the same shear strength as a bolt of the same size. Shank tolerances are similar to AN hex-head bolts，and the threads are National Fine. Structural screws are available with fillister，flat，or washer heads.

（3）There are four types of heads available on self-tapping screws：a round head，a truss head，a countersunk head，which is flat on top，and the countersunk oval screw. The truss-head is rounded，similar to the round head screw，but is considerably thinner.

（4）Washers provide a bearing surface area for nuts，and act as spacers or shims to obtain the proper grip length for a bolt and nut assembly. They are also used to adjust the position of castellated nuts with respect to drilled cotter pin holes in bolts as well as apply tension between a nut and a material surface to prevent the nut from vibrating loose.

(5)Turnlock fasteners are used to secure inspection plates, doors, cowlings, and other removable panels on aircraft. The most desirable feature of these fasteners is that they permit quick and easy removal of access panels for inspection and servicing purposes.

(6)In addition to the numerous permanent fasteners used on aircraft, there is a second type of fastener that, unlike rivets, can be reused. These fasteners include threaded hardware such as bolts and screws and the various types of nuts that secure them.

(7)Drilled-head engine bolts have a thicker head that is drilled with a small hole in each of the flats and in the center of the head. An advantage of drilled-head engine bolts is that they are made with either fine or course threads.

(8)When an internal wrenching bolt is installed in an aluminum alloy structure, a MS20002C washer must be used under the head to provide the needed bearing area. The strength of internal wrenching bolts is much higher than that of a standard steel AN bolt and, for this reason, an AN bolt must never be substituted for an internal wrenching type.

(9)Unless otherwise specified in an assembly drawing, bolts should be installed with their head on top or forward. Placing the head in either of these positions makes it less likely that a bolt will fall out of a hole if the nut is lost. An acronym to help remember the proper direction for bolt installation is "IDA", which stands for inboard, down, or aft.

(10)Many nuts are held on a bolt by passing a cotter pin through a hole in the bolt shank and through slots, or castellations, in the nut.

(11)This insert has a hole slightly smaller than the thread diameter on which it fits. This creates friction between the threads and nut to keep the nut from vibrating loose.

(12)Many special fasteners have the advantage of producing high strength with light weight and can be used in place of conventional AN bolts and nuts.

(13) Lockbolts are available for both shear and tension applications. With shear lockbolts, the head is kept thin and there are only two grooves provided for the locking collar. However, with tension lockbolts, the head is thicker and four or five grooves are provided to allow for higher tension values.

(14)The Hi-Lite fastener is similar to the Hi-Lok except that it is made from lighter materials and has a shorter transition from the threaded section to the shank. Furthermore, the elimination of material between the threads and shank give an additional weight saving with no loss of strength. The Hi-Lite's main advantage is its excellent strength to weight ratio.

(15)The strongest special fastener, used in the construction of many standard and wide-body jet aircraft, is the Taper-Lok, a breakthrough in jet age fasteners. The Taper-Lok was developed to improve the structural fatigue life of modern jet aircraft.

(16)Hi-Tigue fasteners are produced in aluminum, titanium, and stainless steel alloys.

The collars are also composed of compatible metal alloys and are available in two types, sealing and non-sealing.

Ⅲ. Fill in the following blanks according to the text.

(1) Thread fastener include threaded hardware such as _____ and _____ and the various types of _____ that secure them. The most common unthreaded fastener are different kinds of _____.

(2) An acronym to help remember the proper direction for bolt installation is "IDA", which stands for _____, _____, _____.

(3) There are three basic classifications of screws used in aircraft construction: _____, which are the most widely used; _____, which have the same strength as bolts; and _____, which are typically used to join light weight materials.

(4) The two general types of self-locking nuts used in aviation are the _____ type, and the _____ type.

(5) The three most common types of washers used in airframe repair are _____, _____, _____.

EXTENSIVE READING

Designation Codes

The FAA forbids the use of aluminum alloy bolts and alloy steel bolts smaller than AN3 on structural components. Furthermore, since repeated tightening and loosening of aluminum alloy bolts eventually ruins their threads, they are not used in areas where they must be removed and installed frequently. Aluminum alloy nuts can be used with cadmium-plated steel bolts loaded in shear, but only on land aircraft. However, since exposure to moist air increases the possibility of dissimilar-metal corrosion, they cannot be used on seaplanes.

Like rivets, threaded fasteners are given a part code indicating a fastener's diameter in 1/16 inch increments and its length in 1/8 inch increments. For example, an AN4-7 identifies a belt that measures 4/16 or 1/4 inch in diameter and 7/8 inch in length.

For bolts that are longer than 7/8 inch, the code changes. For example, a 1 inch belt is identified by a −10 representing 1 inch and no fraction. In other words, there are no −8 or −9 lengths. Dash numbers go from −7 to −10, from −17 to −20, and from −27 to −30. Therefore, a belt that is 1 1/2 inches long is identified by a −14. A belt with the code AN5−22 identifies an Air Force−Navy belt that is 5/16 inch in diameter and 2 1/4 inches long(Figure 9 − 23, Figure 9 − 24).

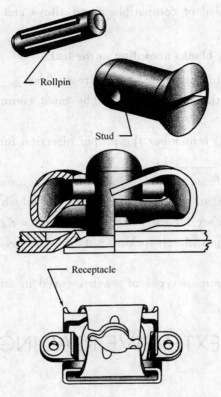

Figure 9 - 23　Airloc cowling fasteners are similar to Dzus fasteners and are used in many of the
same applications

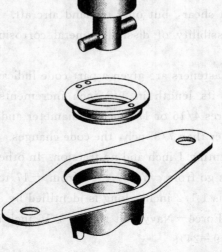

Figure 9 - 24　With a Camlock cowling fastener the stud assembly can be inserted into the
receptacle when the pin is aligned with the slot in the receptacle

Threaded aircraft fasteners 1/4 inch in diameter and smaller are dimensioned in screw sizes rather than 1/8 inch increments. The AN3 belt is the exception to this rule. These machine screw sizes range from 0 to 12. A number 10 fastener has a diameter of approximately 3/16 inch and a number 5 fastener has a 1/8 inch diameter.

When hardware was first standardized, almost all nuts were locked onto a bolt with a cotter pin and, therefore, all bolts had holes drilled near the end of their shank to accommodate a cotter pin. However, when self-locking nuts became popular, many standard AN bolts were made without a drilled shank. To help you identify whether or not a bolt has a hole drilled through it, the letter A is used in the part code. For example, if an "A" appears immediately after the dash number the bolt does not have a hole. However, the absence of an "A" indicates a hole exists in the shank. As an example, an AN6C-12A bolt is 3/8 inch in diameter, made of corrosion-resistant steel, I 1/4 inches long, and has an undrilled shank.

Some AN bolts, such as those used to fasten a propeller into a flanged shaft, must be safetied by passing safety wire through holes drilled through the bolt's head. A bolt drilled for this type of safetying has the letter H following the number indicating its diameter. For example, the part number AN6H34A identifies a bolt that is 3/8 inch in diameter, made of nickel-steel, has a drilled head, is 3 1/2 inches long, and has an undrilled shank.

Lesson 10   Rivets

INTENSIVE READING

SOLID SHANK RIVETS

The solid shank rivet has been used since sheet metal was first utilized in aircraft, and remains the single most commonly used aircraft fastener today. Solid shank rivets are available in a variety of materials, head designs, and sizes to accommodate different applications. Rivets are identified by the shape of the manufactured head, the marking on the head, the rivet part number, and the size.

RIVET CODES

Rivets are given part codes that indicate their size, head style, and alloy material. Two systems are in use today: the Air Force-Navy, or AN system; and the Military Standards 20 system, or MS20. While there are minor differences between the two systems, both use the same method for describing rivets. As an example, consider the rivet designation, AN470AD4 – 5 (Figure 10 – 1).

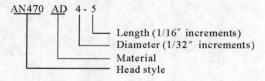

Figure 10 – 1   Rivet identification numbers indicate

The first component of a rivet part number denotes the numbering system used. As discussed, this can either be AN or MS20. The second part of the code is a three-digit number that describes the style of rivet head. The two most common rivet head styles are the universal head, which is represented by the code 470, and the countersunk head, which is represented by the code 426. Following the head designation is a one- or two-digit letter code representing the alloy material used in the rivet. These codes will be discussed in detail later.

After the alloy code, the shank diameter is indicatedin 1/32 inch increments, and the length in increments of 1/16 inch. Therefore, in this example, the rivet has a diameter of 4/32 inch and is 5/16 of an inch long.

The length of a universal head (AN470) rivet is measured from the bottom of the manufactured head to the end of the shank. However, the length of a countersunk rivet (AN426) is measured from the top of the manufactured head to the end of the shank (Figure 10 - 2). Universal and countersunk rivet diameters are measured in the same way, but their length measurements correspond to their grip length.

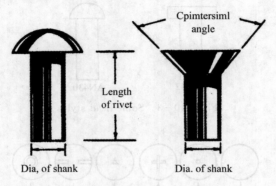

Figure 10 - 2　Methods of measuring rivets

RIVET HEAD DESIGN

As mentioned, solid shank rivets are available in two standard head styles, universal and countersunk, or flush. The AN470 universal head rivet now replaces all previous protruding head styles such as AN430 round, AN442 flat, AN455 brazier, and AN456 modified brazier (Figure 10 - 3). The round head rivet (AN430) was used extensively on aircraft built before 1955, while the flat head rivet (AN442) was widely used on internal structures. Flat head rivets are still used for applications requiring higher head strength.

AN426 countersunk rivets were developed to streamline airfoils and permit a smooth flow over an aircraft's wings or control surfaces.

RIVET ALLOYS

Most aircraft rivets are made of aluminum alloy. The type of alloy is identified by a letter in the rivet code, and by a mark on the rivet head itself (Figure 10 - 4).

1100 Aluminum (A)

Rivets made of pure aluminum have no identifying marks on their manufactured head, and are designated by the letter A in the rivet code. Since this type of rivet is made out of commercially pure aluminum, the rivet lacks sufficient strength for structural applications. Instead, 1,100 rivets are restricted to nonstructural assemblies such as fairings, engine baffles, and furnishings. The 1,100 rivet is driven cold, and therefore, its shear strength increases slightly as a result of cold-working.

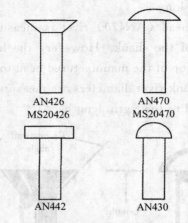

AN426
MS20426

AN470
MS20470

AN442

AN430

Figure10 - 3　Some head styles

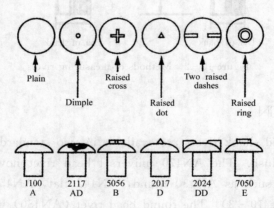

Plain

Dimple

Raised cross

Raised dot

Two raised dashes

Raised ring

1100
A

2117
AD

5056
B

2017
D

2024
DD

7050
E

Figure 10 - 4　Head markings indicate the alloy used in common aircraft rivets

2117 Aluminum Alloy (AD)

The rivet alloy 2117 - T3 is the most widely used for manufacturing and maintenance of modern aircraft. Rivets made of this alloy have a dimple in the center of the head and are represented by the letters AD in rivet part codes. Because AD rivets are so common and require no heat treatment, they are often referred to as "field rivets".

The main advantage for using 2117 - T3 for rivets is its high strength and shock resistance characteristics. The alloy 2117 - T3 is classified as a heat-treated aluminum alloy, but does not require reheat-treat-ment before driving.

5056 Aluminum Alloy (B)

Some aircraft parts are made of magnesium. If aluminum rivets were used on these parts, dissimilar metal corrosion could result. For this reason, magnesium structures are riveted with 5056 rivet. These rivets are identified by a raised cross on their heads and the letter B in a rivet code.

2017 Aluminum Alloy (D)

2017 aluminum alloy is extremely hard. Rivets made of this alloy are often referred to as D rivets, and have been widely used for aircraft construction for many years. D-rivets are identified by a raised dot in the center of their head and the letter D in rivet codes. Because D-rivets are so hard, they must be heat-treated before they can be used (Figure 10 – 4).

Recall from the study of heat treatments, that when aluminum alloy is quenched after heat treatment, it does not harden immediately. Instead, it remains soft for several hours and gradually becomes hard and gains full strength. Rivets made of 2017 can be kept in this annealed condition by removing them from a quench bath and immediately storing them in a freezer. Because of this, D-rivets are often referred to as icebox rivets. These rivets become hard when they warm up to room temperature, and may be reheat-treated as many times as necessary without impairing their strength.

2024 Aluminum Alloy (DD)

DD-rivets are identified by two raised dashes on their head. Like D-rivets, DD-rivets are also called icebox rivets and must be stored at cool temperatures until they are ready to be driven. When DD-rivets are driven, their alloy designation becomes 2024 – T31 because of the work hardening achieved during installation.

Of the two icebox rivets, the 2024 – T4 rivet has the greatest bearing and shear strength. The two icebox rivets are used in areas of the aircraft where more than the usual amount of strength is required.

7050 – T73 Aluminum Alloy (E)

A new and stronger rivet alloy was developed in 1979 called 7050 – T73. The letter E is used to designate this alloy, and the rivet head is marked with a raised circle. 7050 alloy contains zinc as the major alloying ingredient, and is precipitation heat-treated. This alloy is used by the Boeing Company as a replacement for 2024 – T31 rivets in the manufacture of the 767 wide-body aircraft.

Corrosion-Resistant Steel (F)

Stainless steel rivets are used for fastening corrosion-resistant steel sheets in applications such as firewalls and exhaust shrouds. They have no marking on their heads.

Monel（M）

Monel rivets are identified with two recessed dimples in their heads. They are used in place of corrosion-resistant steel rivets when their somewhat lower shear strength is not a detriment.

New Words/Phrases/Expression

1. sheet metal　金属薄板、钣金
2. Air Force-Navy（AN）　美国空军海军标准
3. Military Standard（MS）　军用标准
4. part number　件号
5. universal[juːnɪ'vɜːs(ə)l]　adj.普遍的；通用的
6. countersunk['kauntəsʌŋk]　n.&adj.埋头孔，埋头的
7. grip length　夹紧长度
8. protruding[prə'truːdɪŋ]　adj.突出的；伸出的
9. streamline['striːmlaɪn]　adj.流线型的
10. airfoil['eəfɔɪl]　n.翼面；机翼；螺旋桨
11. baffle['bæf(ə)l]　n.挡板
12. bearing strength　承压强度
13. shear strength　剪切强度
14. cold-working　冷加工
15. field rivet　外场铆钉
16. shock resistance　耐冲击性
17. dissimilar metal corrosion　异类金属腐蚀
18. icebox rivet　冰盒铆钉
19. firewall['faiəwɔːl]　n.防火墙
20. exhaust[ɪg'zɔːst; eg-]　n.排气；废气；排气装置
21. shroud[ʃraud]　n.覆盖物；[电]护罩

Notes

（1）The two most common rivet head styles are the universal head，which is represented by the code 470，and the countersunk head，which is represented by the code 426.

分析：由 which 引导的非限定性定语从句；be represented by 用……表示。

翻译：通用头和埋头是最常用的两种铆钉头类型。通用头用代码 470 表示，埋头用代码 426 表示。

（2）Universal and countersunk rivet diameters are measured in the same way，but their length measurements correspond to their grip length.

分析：correspond to 对应于……，与……相对应。

翻译：通用铆钉和埋头铆钉的直径是用相同的方法进行测量，而它们的长度测量是与之夹紧长度相对应。

（3）After the alloy code, the shank diameter is indicated in 1/32 inch increments, and the length in increments of 1/16 inch.

分析：and 后面是并列句，省略了谓语动词。

翻译：在表示合金材料的代码后面是铆钉杆直径，直径的每个单位为 1/32 英寸，长度的每个单位为 1/16 英寸。

（4）Rivets made of this alloy have a dimple in the center of the head and are represented by the letters AD in rivet part codes.

分析：made of this alloy 是定语从句。

翻译：这种合金制成的铆钉，在它头部中心位置有一个凹点，在铆钉件号中用字母代码AD 表示。

Exercises

Ⅰ. Fill in the right information in the two forms.

Head marking	Letter	Alloy
Plain	A	1100
Dimple		
Raised cross		
Raised dot		
Two raised dashes		
Raised ring		

Rivet			
Code number	AN426		
Head style	countersunk head		

Ⅱ. Fill in the following blanks according to the text.

（1）Rivets are identified by _____ , _____ , and _____ .

（2）Rivet codes indicate _____ .

（3）AN470AD5 - 6indicates that its head style is _____ .

（4）AN426D3 - 5 indicates that its length is _____ inch.

(5) Which rivet now replaces almost all other protruding head designs? _____
_____.

(6) Which rivet is identified by a rasised dot on the rived head? _____.

(7) Which rivet is the field rivets? _____.

(8) Which rivet is commonly called icebox revit? _____.

(9) The length of an _____ rivet is measured from the top of the manufactured head to the end of the shank.

(10) All aluminum alloy universal head rivets will have the _____ code number.

(11) The _____ rivet is used where a smooth surface is of greatest importance.

(12) The AN442 rivet is _____. The rivet was widely used on internal structures.

(13) The _____ aluminum alloy rivet is generally used when riveting magnesium parts.

(14) _____ require special treatment.

(15) Of the two icebox rivets, the _____ rivet has the greatest bearing and shear strength.

(16) The _____ rivets are used in areas of the aircraft where more than the usual amount of strength is required.

(17) Universal and countersunk rivet diameters are measured in the same way, but their length measurements correspond to _____.

Ⅲ. **Answer the following questions.**

(1) How many tanks are there in the fuel system?

(2) How to measure the length of a countersunk rivet or universal rivet?

(3) Where the icebox rivets are used in aircraft? Why?

(4) Describe the two basic rivet head styles.

(5) What is the field rivets?

(6) What's the meaning of AN470AD6 – 8, MS20426DD5 – 7?

Ⅳ. **Translate the following sentences into Chinese.**

(1) Rivets are identified by the shape of the manufactured head, the marking on the head, the rivet part number, and the size.

(2) The two most common rivet head styles are the universal head, which is represented by the code 470, and the countersunk head, which is represented by the code 426.

(3) After the alloy code, the shank diameter is indicated in 1/32 inch increments, and the length in increments of 1/16 inch.

(4) Universal and countersunk rivet diameters are measured in the same way, but their length measurements correspond to their grip length.

(5) The type of alloy is identified by a letter in the rivet code, and by a mark on the rivet head itself.

(6) The alloy 2117 – T3 is classified as a heat-treated aluminum alloy, but does not require reheat-treat-ment before driving.

(7) Of the two icebox rivets, the 2024 – T4 rivet has the greatest bearing and shear strength.

EXTENSIVE READING

Special Rivets

Conventional solid shank rivets require access to both ends to be driven. However, special rivets, often called blind rivets are installed with access to only one end of the rivet. While considerably more expensive than solid shank rivets, blind rivets find many applications in today's aircraft industry.

Pop™ Rivets

Pop rivets have limited use on aircraft and are never used for structural repairs. However, they are useful for temporarily lining up holes. In addition, some "home built" aircraft utilize Pop rivets. They are available in flat head, countersunk head, and modified flush heads with standard diameters of 1/8, 5/32, and 3/16 inch. Pop rivets are made from soft aluminum alloy, steel, copper, and Monel (Figure 10 – 5).

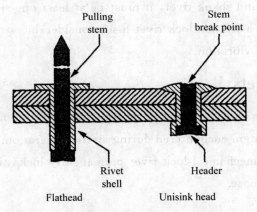

Figure 10 – 5   Pop rivets are frequently used for assembly and non — structural applications. They must not be used in areas that are subject to moderate or heavy loads

Friction-lock rivets

One early form of blind rivet that was the first to be widely used for aircraft construction and repair was the Cherry friction-lock rivet. Originally, Cherry friction-locks were available in two styles: hollow shank pull-through, and self-plugging types. The pull-through type is no longer common. However, the self-plugging Cherry friction-lock rivet is still used for repairing light aircraft.

Cherry friction-lock rivets are available in two head styles: universal and 100 degree countersunk. Furthermore, they are usually supplied in three standard diameters: 1/8, 5/32, and 3/16 inch. However, larger sizes can be specially ordered in sizes up to 5/16 inch (Figure 10 – 6).

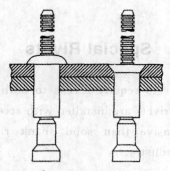

Figure 10 – 6　The friction-lock rivet assembly consists of a shell and mandrel, or pulling stem. The stem is pulled until the header forms a buck-tail on the blind side of the shell. At this point, a weak point built into the stem shears and the stem breaks off. After the stem fractures, part of it projects upward. The projecting stem is cut close to the rivet head, and the small residual portion remaining in the head is filed smooth

A friction-lock rivet cannot replace a solid shank rivet, size for size. When a friction-lock is used to replace a solid shank rivet, it must be at least one size (1/32 inch) larger in diameter. This is because a friction-lock rivet loses considerable strength if its center stem falls out due to damage or vibration.

Mechanical-lock rivets

Mechanical-lock rivets were designed to prevent the center stem of a rivet from falling out as a result of the vibration encountered during aircraft operation. Unlike the center stem of a friction-lock rivet, a mechanical-lock rivet permanently locks the stem into place, and vibration cannot shake it loose.

Huck-loks

Huck-Lok rivets were the first mechanical-lock rivets and are used as structural replacements for solid shank rivets. However, because of the expensive tooling required for their installation, Huck-Loks are generally limited to aircraft manufacturers and some large repair facilities.

Huck-Loks are available in four standard diameters: 1/8, 5/32, 3/16, and 1/4 inch, and come in three different alloy combinations: a 5056 sleeve with a 2024 pin, an A-286 sleeve with an A-286 pin, and a Monel 400 sleeve with an A-286 pin (Figure 10 – 7).

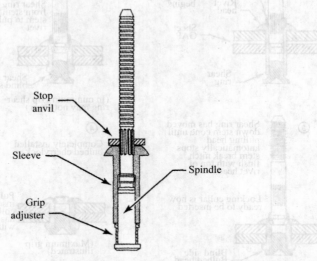

Figure 10 – 7　Unlike friction-lock rivets, Huck-Loks utilize a lock ring that mechanically locks the center stem in place

CherryLOCKs™

The Cherry mechanical-lock rivet, often called the bulbed CherryLOCK, is an improvement over the friction-lock rivet because its center stem is locked into place with a lock ring. This results in shear and bearing strengths that are high enough to allow CherryLOCKS to be used as replacements for solid shank rivets (Figure 10 – 8).

CherryLOCK rivets are available with two head styles, 100 degree countersunk and universal. Like most blind rivets, CherryLOCKs are available with diameters of 1/8, 5/32, and 3/16 inch, with an oversize of 1/64 inch for each standard size. The rivet, or shell, portion of a CherryLOCK may be constructed of 2017 aluminum alloy, 5056 aluminum alloy, Monel, or stainless steel. Installation of CherryLOCK rivets requires a special pulling tool for each different size and head shape. However, the same size tool can be used for an oversize rivet in the same diameter group. One disadvantage of a CherryLOCK is that if a rivet is too short for an application, the lock ring sets prematurely resulting in a malformed shank header. This fails to compress the joint, leaving it in a weakened condition. To avoid this, always use the proper rivet length selection gauge and follow the manufacturer's installation recommendations.

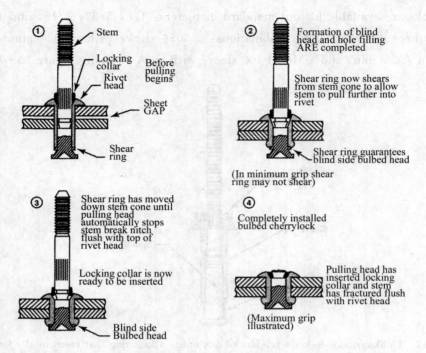

Figure 10 – 8　(1) As the stem is pulled into the rivet sleeve, a bulb forms on the rivet's blind side that begins to clamp the two pieces of metal together and fill the hole. (2) Once the pieces are clamped tightly together, the bulb continues to form until the shear ring shears and allows the stem to pull further into the rivet. (3) With the shear ring gone, the stem is pulled upward until the pulling head automatically stops at the stem break notch, and the locking collar is ready to be inserted. (4) When completely installed, the locking collar is inserted and the stem is fractured flush with the rivet head

Olympic-loks

Olympic-lok blind fasteners are lightweight, mechanically-locking, spindle-type blind rivets. Olympic-loks come with a lock ring stowed on the head. As an Olympic-lok is installed, the ring slips down the stem and locks the center stem to the outer shell. These blind fasteners require a specially designed set of installation tools (Figure 10 – 9).

Olympic-lok rivets are made with three head styles: universal, 100 degree flush, and 100 degree flush shear. Rivet diameters of 1/8, 5/32, and 3/16 inch are available in eight different alloy combinations of 2017 – T4, A – 286, 5056, and Monel.

When Olympic-loks were first introduced, they were advertised as an inexpensive blind fastening system. The price of each rivet is less than the other types of mechanical locking blind rivets, and only three installation tools are required. The installation tools fit both countersunk and universal heads in the same size range.

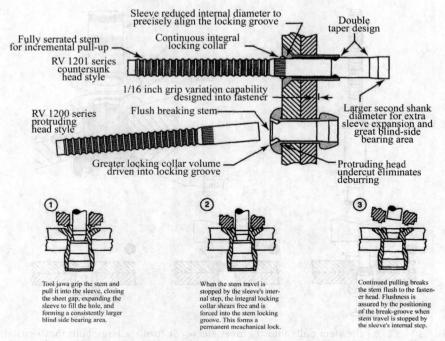

Sleeve reduced internal diameter to precisely align the locking groove

Double taper design

Fully serrated stem for incremental pull-up

Continuous integral locking collar

RV 1201 series countersunk head style

1/16 inch grip variation capability designed into fastener

Larger second shank diameter for extra sleeve expansion and great blind-side bearing area

RV 1200 series protruding head style

Flush breaking stem

Greater locking collar volume driven into locking groove

Protruding head undercut eliminates deburring

① Tool jaws grip the stem and pull it into the sleeve, closing the sheet gap, expanding the sleeve to fill the hole, and forming a consistently larger blind side bearing area.

② When the stem travel is stopped by the sleeve's internal step, the integral locking collar shears free and is forced into the stem locking groove. This forms a permanent mechanical lock.

③ Continued pulling breaks the stem flush to the fastener head. Flushness is assured by the positioning of the break-groove when stem travel is stopped by the sleeve's internal step.

Figure 10 – 9　(1) Once an Olympic-Lok rivet is inserted into a prepared hole, the stem is pulled into the sleeve, closing any gap between the materials being riveted; filling the hole, and forming a bearing area. (2) When the stem travel is stopped by the sleeve's internal step, the locking collar shears free and is forced into the locking groove. (3) Continued pulling breaks the stem flush with the rivet head

CherryMAX™

The CherryMAX rivet is economical to use and strong enough to replace solid shank rivets, size for size. The economic advantage of the CherryMAX system is that one size puller can be used for the installation of all sizes of CherryMAX rivets. A CherryMAX rivet is composed of five main parts: a pulling stem, a driving anvil, a safe-lock locking collar, a rivet sleeve, and a bulbed blind head (Figure 10 – 10).

Available in both universal and countersunk head styles, the rivet sleeve is made from 5056, monel, and inco 600. The stems are made from alloy steel, CRES, and inco X – 750. The ultimate shear strength of CherryMAX rivets ranges from 50KSI to 75KSI. Furthermore, CherryMAX rivets can be used at temperatures from 250℉ to 1,400℉. They are available in diameters of 1/8, 5/32, 3/16 and 1/4 inches, and are also made with an oversize diameter for each standard diameter listed.

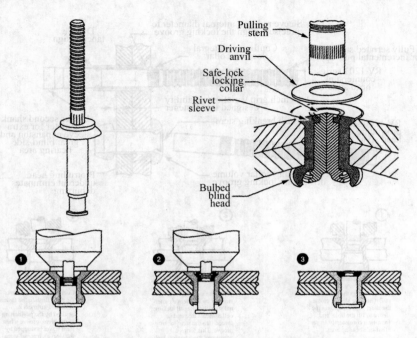

Figure 10 - 10　(1) As the stem pulls into the rivet sleeve, it forms a large bulb that seats the rivet head, and clamps the two sheets tightly together. (2) As the blind head is completed, the safe-locking collar moves into the rivet sleeve recess. (3) As the stem continues to be pulled, the safe-lock collar is formed into the head recess by the driving anvil, locking the stem and sleeve securely together. Further pulling fractures the stem, providing a flush, burr-free installation.

Removal of Mechanical-Lock Rivets

To remove mechanical-lock rivets, first file a flat spot on the rivet's center stem. Once this is done, a center punch is used to punch out the stem so the lock ring can be drilled out. With the lock ring removed, tap out the remaining stem, drill to the depth of the manufactured head, and tap out the remaining shank. All brands of mechanical-lock blind rivets are removed using the same basic technique.

Hi-Shear rivets

One of the first special fasteners used by the aerospace industry was the Hi-Shear rivet. Hi-Shear rivets were developed in the 1940s to meet the demand for fasteners that could carry greater shear loads.

The Hi-Shear rivet has the same strength characteristics as a standard AN bolt. In fact, the only difference between the two is that a bolt is secured by a nut and a Hi-Shear rivet is secured by a crushed collar. The Hi-Shear rivet is installed with an interference fit, where the side wall clearance is reamed to a tolerance determined by the aircraft builder. When properly installed, a Hi-Shear rivet has to be tapped into its hole before the locking collar is swaged on.

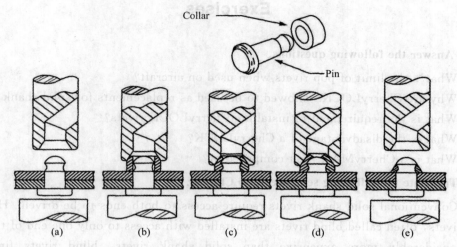

Figure 10 – 11.　(a) A bucking bar and rivet gun are used to install Hi-Shear rivets.　(b)
A collar is placed over the pin's small end.　(c) The rivet gun forces the
collar over the pin.　(d) The gun set drives the collar onto the rivet pin
and cuts off excess material.　(e) When the collar is fully driven, excess
collar material is ejected from the gun set.

　　Hi-Shear rivets are made in two head styles: flat and countersunk. As the name
implies, the Hi-Shear rivet is designed especially to absorb high shear loads. The Hi-Shear
rivet is made from steel alloy having the same tensile strength as an equal size AN bolt. The
lower portion of its shank has a specially milled groove with a sharp edge that retains and
finishes the collar as it is swaged into the locked position (Figure 10 – 11).

New Words/Phrases/ Expression

1. conventional[kən'venʃ(ə)n(ə)l]　*adj.* 传统的；常见的；惯例的
2. application[æplɪ'keɪʃ(ə)n]　*n.* 应用；申请；应用程序
3. temporarily['temp(ə)r(ər)ɪlɪ]　*adv.* 临时地
4. available[ə'veɪləb(ə)l]　*adj.* 可获得的；可找到的
5. bulbed[bʌlbd]　*adj.* 球状的，圆头的
6. characteristic[kærəktə'rɪstɪk]　*n.* 特征；特性；特色
7. interference[ɪntə'fɪər(ə)ns]　*n.* 干扰，冲突；干涉
8. clearance['klɪər(ə)ns]　*n.* 清除；空隙；间隙
9. ream[riːm]　*vt.* 扩展；挖；铰孔
10. milled[mɪld]　*adj.* 磨碎的；滚花的；铣成的
11. groove[gruːv]　*n.* 凹槽，槽
12. collar['kɒlə]　*n.* 衬套；颈圈

Exercises

Ⅰ. **Answer the following questions.**

(1) What's the limit of pop rivets when used on aircraft?

(2) Why are CherryLOCKS allowed to be used as replacements for solid shank rivets?

(3) What is the requirement of installing CherryLOCK rivets?

(4) What is the disadvantage of a CherryLOCK?

(5) What is a CherryMAX rivet composed of?

Ⅱ. **Translate the following sentences into Chinese.**

(1) Conventional solid shank rivets require access to both ends to be driven. However, special rivets, often called blind rivets are installed with access to only one end of the rivet. While considerably more expensive than solid shank rivets, blind rivets find many applications in today's aircraft industry.

(2) The CherryMAX rivet is economical to use and strong enough to replace solid shank rivets, size for size. The economic advantage of the CherryMAX system is that one size puller can be used for the installation of all sizes of CherryMAX rivets.

(3) In fact, the only difference between the two is that a bolt is secured by a nut and a Hi-Shear rivet is secured by a crushed collar. The Hi-Shear rivet is installed with an interference fit, where the side wall clearance is reamed to a tolerance determined by the aircraft builder.

(4) The Hi-Shear rivet is made from steel alloy having the same tensile strength as an equal size AN bolt.

Lesson 11　Installation of Rivets

INTENSIVE READING

INSTALLATION OF SOLID RIVETS

A primary task performed by any aircraft maintenance technician working with sheet metal is to install solid shank rivets in structures of components. This involves selecting the proper rivets and installing them in such a way that the maximum structural integrity and aerodynamic shape are attained. Although the concept of installing rivets is a straightforward process, there are a number of aspects that the technician must understand and observe to achieve the optimum performance from the structure or component design. One of the most critical aspects of sheet metal work is to be able to select the proper rivet for a given application.

In a general formula, the proper diameter of rivet is equal to three times the thickness of the thickest sheet of metal in the joint. Once the proper diameter of rivet has been determined, the correct rivet length can be calculated.

Rivet Layout Patterns

It is important when making a riveted repair that the rivets beinstalled in such a way that they will develop the maximum strength from the sheet metal. To obtain this strength, not only the rivet and sheet strength must be determined, but the rivet pattern is also a critical factor so the drilled holes do not weaken the joint. This means the spacing between rivets and the distance they remain from the edge of the material cannot be closer than minimum specifications.

If rivets areinstalled too close to the edge, the sheet metal will tear out instead of shearing the rivet when extreme loads are encountered. Conversely, if the rivets are placed too far away from the edge, the metal sheets can separate, allowing foreign contaminates to enter the joint, ultimately causing corrosion.

An accepted practice is to place the center of a rivet hole nocloser than two rivet shank diameters from the edge and no further back than four diameters.

Pitch

The distance between adjacent rivet in a row is called the pitch. To prevent the joint from being weakened by too many holes in a row, theadjacent rivets should be no closer than three diameters to one another. In contrast, to prevent the sheets from separating between rivets, the rivet holes should be no further apart than ten to twelve times the rivet shank diameter (Figure 11 – 1).

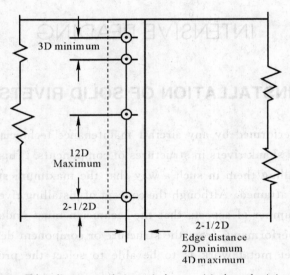

Figure 11 – 1   Edge distance and rivet pitch are critical to obtaining maximum
strength from a riveted repair

Rivet Gauge or Transverse Pitch

The distance between rows of rivets in a multi-row layout should be about 75% of the pitch, provided that the rivets in adjacent rows are staggered. If the rivets are not staggered, then the pitch will be the same between rows as it is between rivets in a single row.

The holes made by these twist drills are usually three- or four-thousandths of an inch larger than the diameter of the rivet. This allows the rivet to be slipped in place without forcing it and scraping any protective oxide coating off the rivet shank.

Deburring

When aluminum alloys are drilled, sharp burrs can remain on the edge of the hole. To remove these sharp edges and burrs, a process referred to as deburring is accomplished.

Hole Preparation for Flush Rivets

However, before a countersunk rivet can be installed, the metal must be countersunk or dimpled. Countersinking is a process in which the metal in the top sheet is cut away in the shape of the rivet head. In some situations, the surface material is machined away by a countersink cutter. On the other hand, dimpling is a process that mechanically "dents" the sheets being joined to accommodate the rivet head. Sheet thickness and rivet size determine which method is best suited for a particular application.

Rivet Installation

In addition to the proper preparation of a hole for a rivet, the strength of a riveted joint is determined by the way the rivets are driven. When installing rivets, it is important to install the rivet with as few impacts as possible so the materials will not work-harden and crack. The shop head of the rivet should be concentric with the shank and flush with the surface without tipping. In addition, the formed, or bucked head should be fabricated to proper dimensions (Refer to Figure 11 – 3).

Compression Riveting

A rivet gun drives most rivets used in aircraft construction. When there are a large number of easily accessible rivets to be installed, a compression, or squeeze riveter, can be used instead of gun riveting. These riveting tools reduce the time required to install the rivets and produce a far more uniform shape than can be driven with a rivet gun.

Rivet Sets

Rivet sets are manufactured differently for various head styles and sizes of rivets. When selecting rivet sets, the radius of the depression in the set must be larger than that of the rivet, but not so large that the set contacts the sheet metal to produce small indentations which are commonly referred to as smiles during driving. On the other hand, if the set is too small, it will produce a similar type mark on the rivet head, which is also unacceptable (Figure 11 – 2). In addition, damage on the recess face of the set may cause it to slip off the rivet during driving, or the set may leave unacceptable marks on the rivet head.

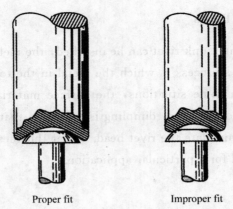

Proper fit　　　　　　Improper fit

Figure 11 - 2　The radius of the cup of the rivet set must be slightly larger than the radius of the rivet head, but not so large that the edges of the set contact the surface of the metal

Bucking Bars

When driven, the rivet is backed up by a metal bar that reciprocates in response to the beats of the rivet gun. This reciprocating action causes the rivet to be compressed in successive actions. These metal bars are referred to as bucking bars because of the method in which the bar bucks, or vibrates, on the shank of the rivet. The driving face of a bucking bar is machined smooth and polished so that no marks are left on the rivet shank during driving. There are many sizes and shapes of bucking bars used in aircraft maintenance. The weight of the bucking bar must be proportional to the size of the rivet.

Rivet Installation

Once the metal has been drilled, put a rivet of the correct length through a hole and hold the rivet gun set against the manufactured head of the rivet. The set must be directly in line with the rivet, and not tipped or it will contact the sheet metal during driving. Hold a bucking bar flat against the end of the rivet shank and develop a good feel of the balance between the gun and the bucking bar. Once the gun is in position, pull the trigger to provide a short burst of raps on the rivet head. The rivet should be driven with the fewest blows possible so it will not work-harden the rivet, which can cause cracks to form.

New Words/Phrases/Expression

1. primary['praɪm(ə)rɪ]　*adj.* 主要的；初级的；基本的
2. maintenance technician　维修技术员
3. structural integrity　结构完整性
4. aerodynamic[ˌɛrodaɪ'næmɪk]　*adj.* 空气动力学的
5. optimum performance　最佳操作特性；最佳性能

6. general formula　通式；一般公式；一般准则

7. layout pattern　布局模式

8. extreme load　极限载荷

9. accepted practice　习惯作法；常例

10. far back　遥远

11. pitch[pɪtʃ]　n. 间距、间隙

12. gauge[geɪdʒ]　n. 行距

13. staggered[ˈstægəd]　adj. 错列的；吃惊的

14. scrape[skreɪp]　vt. 刮；擦伤

15. deburring[diˈbəːrɪŋ]　n. 倒角；去除毛刺；清除飞边

16. countersink cutter　埋头钻

17. work-harden　加工硬化

18. squeeze riveter　压铆机

19. bucking bar　顶铁

Notes

(1) In a general formula, the proper diameter of rivet is equal to three times the thickness of the thickest sheet of metal in the joint.

分析：句型"be equal to..."属于等比句型，比较双方基本等同，标准形式是"主语十 be equal to十等比对象"。形容词"equal"的意思是"相等的、同样的"。该句型翻译为"……和……相当"。

翻译：在通用公式中，合适的铆钉直径应等于所连接的钣金件中最厚板的厚度的三倍。

(2) To obtain this strength, not only the rivet and sheet strength must be determined, but the rivet pattern is also a critical factor so the drilled holes do not weaken the joint.

分析：not only ... but also ... 在句中常用来连接两个对等的成分，表示"不仅……而且……"

翻译：为了获得这个强度，不仅铆钉和板的强度必须被确定，而且为了使所钻的孔不会削弱连接，铆钉的布局也是关键因素。

(3) The distance between rows of rivets in a multi-row layout should be about 75% of the pitch, provided that the rivets in adjacent rows are staggered.

分析：provided that 表示"如果；假如"。

翻译：假如相邻的两行铆钉是交错排布，多行布局的铆钉行距应该是间距的 75%。

(4) When installing rivets, it is important to install the rivet with as few impacts as possible so the materials will not work-harden and crack.

分析：as ... as possible 结构中，表示"尽可能地……"。

翻译：当安装铆钉时，用尽可能少的冲击次数来安装铆钉，这样，材料就不会加工硬化和开裂。

(5) The set must be directly in line with the rivet, and not tipped or it will contact the

sheet metal during driving.

分析:in line with 表示"和······成一直线"。

翻译:铆窝必须与铆钉恰好成一直线,不能倾斜,否则,打铆过程中,铆窝会接触到钣金件。

Exercises

Ⅰ. Fill in the following blanks according to the text.

(1) In a general formula, the proper diameter of rivet is equal to _____ times the thickness of the _____ sheet of metal in the joint.

(2) If rivets are installed too close to the edge, the sheet metal will _____ instead of _____ the rivet when _____ loads are encountered. Conversely, if the rivets are placed too _____ from the edge, the metal sheets can separate, allowing foreign contaminates to enter the joint, ultimately causing _____.

(3) An accepted practice is to place the center of a rivet hole no closer than _____ rivet shank diameters from the edge and no further back than _____ diameters.

(4) The distance between adjacent rivet in a row is called the _____.

(5) To prevent the joint from being weakened by too many holes in a row, the adjacent rivets should be no closer than _____ diameters to one another. In contrast, to prevent the sheets from separating between rivets, the rivet holes should be no further apart than _____ times the rivet shank diameter.

(6) The distance between rows of rivets in a multi-row layout should be about _____ of the pitch, provided that the rivets in adjacent rows are staggered. If the rivets are not staggered, then the pitch will be the _____ between rows as it is between rivets in a single row.

(7) The holes made by these twist drills are usually _____ of an inch _____ than the diameter of the rivet.

(8) To remove these sharp edges and burrs, a process referred to as _____ is accomplished.

(9) Before a countersunk rivet can be installed, the metal must be _____.

(10) _____ determine which method is best suited for a particular application.

(11) When there are a large number of easily accessible rivets to be installed, a _____, can be used instead of gun riveting.

(12) These metal bars are referred to as _____ because of the method in which the bar bucks, or vibrates, on the shank of the rivet.

(13) The driving face of a bucking bar is machined _____ and _____ so that no marks are left on the rivet shank during driving.

(14) The weight of the bucking bar must be proportional to the _____ of the

rivet.

II. Answer the following questions.

(1) What is one of the most critical aspects of sheet metal work?

(2) How to determine the proper diameter river in the general formula?

(3) What will happen if rivets are installed too close to or too far away from the edge?

(4) What should the distance between rows of rivets in a multi-row layout be?

(5) The holes made by these twist drills are usually three- or four-thousandths of an inch larger than the diameter of the rivet, why?

(6) Why do we need the process-deburring?

(7) Which tools can reduce the time required to install the rivets and produce a far more uniform shape than the rivet gun?

III. Translate the following sentences into Chinese.

(1) This involves selecting the proper rivets and installing them in such a way that the maximum structural integrity and aerodynamic shape are attained.

(2) Although the concept of installing rivets is a straightforward process, there are a number of aspects that the technician must understand and observe to achieve the optimum performance from the structure or component design.

(3) To obtain this strength, not only the rivet and sheet strength must be determined, but the rivet pattern is also a critical factor so the drilled holes do not weaken the joint.

(4) The distance between adjacent rivet in a row is called the pitch. To prevent the joint from being weakened by too many holes in a row, the adjacent rivets should be no closer than three diameters to one another. In contrast, to prevent the sheets from separating between rivets, the rivet holes should be no further apart than ten to twelve times the rivet shank diameter.

(5) In addition to the proper preparation of a hole for a rivet, the strength of a riveted joint is determined by the way the rivets are driven. When installing rivets, it is important to install the rivet with as few impacts as possible so the materials will not work-harden and crack. The shop head of the rivet should be concentric with the shank and flush with the surface without tipping. In addition, the formed, or bucked head should be fabricated to proper dimensions.

(6) A rivet gun drives most rivets used in aircraft construction. When there are a large number of easily accessible rivets to be installed, a compression, or squeeze riveter, can be used instead of gun riveting. These riveting tools reduce the time required to install the rivets and produce a far more uniform shape than can be driven with a rivet gun.

(7) When selecting rivet sets, the radius of the depression in the set must be larger than that of the rivet, but not so large that the set contacts the sheet metal to produce small indentations which are commonly referred to as smiles during driving. On the other hand, if the set is too small, it will produce a similar type mark on the rivet head, which is also

unacceptable.

(8) Once the metal has been drilled, put a rivet of the correct length through a hole and hold the rivet gun set against the manufactured head of the rivet. The set must be directly in line with the rivet, and not tipped or it will contact the sheet metal during driving. Hold a bucking bar flat against the end of the rivet shank and develop a good feel of the balance between the gun and the bucking bar.

EXTENSIVE READING

Unlike other types of fasteners, rivets change in dimension to fit the size of a hole during installation (Figure 11 – 3). To fabricate a properly driven rivet, the width of the bucked head must be equal to one and a half times the rivet's original shank diameter and the height must be one half the original shank diameter. In order for the bucked head to develop these finished dimensions, the rivet needs to protrude through the metal approximately one and a half times the shank diameter before being driven.

When a rivet is driven, its cross sectional area increases along with its bearing and shearing strengths.

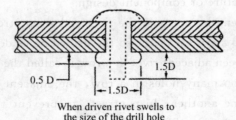

When driven rivet swells to
the size of the drill hole

Figure 11 – 3   Before a rivet is driven, it should extend beyond the base material at least one and a
                 half times the rivet's diameter. Once driven, the rivet shank expands to fill the hole,
                 and the bucktail expands to one and a half times its original diameter. Once the
                 bucktail expands to the appropriate diameter, it should extend beyond the base
                 material by at least one-half the original rivet diameter

Evaluating Driven Rivets

In the process of riveting, it is inevitable that some rivets will be driven improperly. One of the requirements of a technician is to be able to quickly identify rivets with unacceptable characteristics. In addition, the technician must be able to remove damaged or improperly installed rivets without adversely affecting the base sheet metal, and then be able to correct the problem with the installation of replacement rivets (Figure 11 – 4).

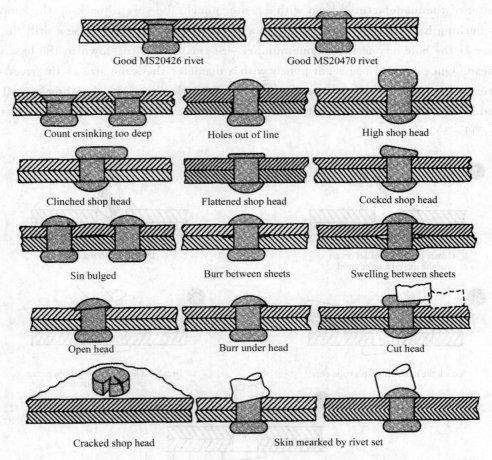

Good MS20426 rivet   Good MS20470 rivet

Count ersinking too deep   Holes out of line   High shop head

Clinched shop head   Flattened shop head   Cocked shop head

Sin bulged   Burr between sheets   Swelling between sheets

Open head   Burr under head   Cut head

Cracked shop head   Skin mearked by rivet set

Figure 11 – 4 A technician must be able to identify properly driven rivets. Until the technician has gained enough experience to evaluate driven head dimensions, special gauges may be fabricated to check shop head shapes and sizes

As previously mentioned, a properly formed shop head will be one-half the shank diameter in height with a diameter that is one and a half times that of the shank diameter. In addition, the rivet should be driven concentric with the hole. Placing a straightedge on top of shop heads that have been driven in a row can check the uniformity of rivet heights. Each surface of the shop heads should touch the straightedge without gaps. The rivet shank should also be checked for concentricity.

The manufactured head of the rivet must also be perfectly flat against the metal. If the hole was improperly drilled, the only satisfactory repair is to re-drill the metal to accept the next larger diameter rivet.

Rivet Removal

When it has been determined that a rivet has been improperly installed, the rivet must be removed and replaced without damaging the base metal. To remove a rivet, lightly indent the center of the manufactured head with a center punch. Be sure to back-up the shop head with a bucking bar when center punching so as not to distort the skin. Use a drill the same diameter as the hole or one that is one number size smaller to drill down to the base of the rivet head. Once drilled, use a pin punch with a diameter the same size as the rivet shank diameter to pry the head off, or tap the head lightly with a cape chisel to break it off from the shank. If a chisel is used, be sure that it does not scratch the skin around the rivet head (Figure 11 - 5).

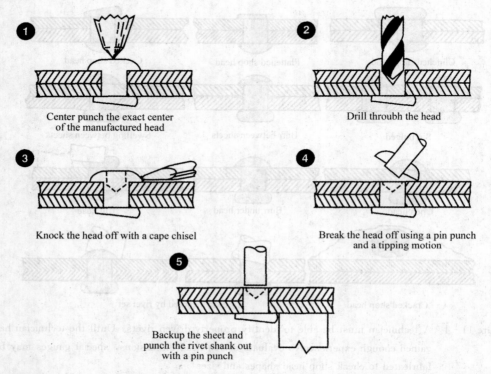

Figure 11 - 5　When removing a solid rivet, it is important to avoid damaging the sheet metal. Although a cape chisel may be used, removing the rivet head with a pin punch is preferred to help avoid damaging the base metal

After the rivet head has been removed, back-up the underside of the skin with a bucking bar or piece of wood, and use a pin punch to gently drive the rivet shank from the sheet metal. When the rivet is out, examine the hole, and if it is not elongated, another rivet of the same size may be used as a replacement. If the hole is damaged, use a twist drill for the next larger diameter rivet to re-drill the hole. When using a larger diameter rivet, be sure that the pitch, gauge, and edge distance values are all satisfactory for the pattern.

New Words/Phrases/Expression

1. evaluate[ɪˈvæljueɪt]　*vt.* 评价；估价

2. fabricate['fæbrɪkeɪt]　*vt.* 制造；装配

3. approximately[ə'prɒksɪmətlɪ]　*adv.* 大约，近似地；近于

4. inevitable[ɪn'evɪtəb(ə)l]　*adj.* 必然的，不可避免的

5. adversely[æd'vɜːslɪ]　*adv.* 不利地；逆地；反对地

6. concentric[kən'sentrɪk]　*adj.* 同轴的；同中心的

7. straightedge['streɪtedʒ]　*n.* (画直线用的)直尺

8. uniformity[juːnɪ'fɔːmɪtɪ]　*n.* 均匀性；一致；同样

9. gap[gæp]　*n.* 间隙；缺口；差距

10. indent[ɪn'dent]　*n.* 缩进；订货单；凹痕；契约

11. back-up['bækʌp]　*v.* 支持，援助

12. distort[dɪ'stɔːt]　*vt.* 扭曲；使失真；曲解

13. pry[praɪ]　*vt.* 撬动，撬开

14. tap[tæp]　*vt.* 轻敲；轻打；装上嘴子

Exercises

Ⅰ. Fill in the following blanks according to the text.

(1) To fabricate a properly driven rivet, the width of the bucked head must be equal to _____ the rivet's original shank diameter and the height must be _____ the original shank diameter.

(2) In order for the bucked head to develop these finished dimensions, the rivet needs to protrude through the metal approximately _____ times the shank diameter before being driven.

(3) If the hole was improperly drilled, the only satisfactory repair is to re-drill the metal to accept _____.

(4) To remove a rivet, lightly indent the center of the manufactured head with a _____.

(5) Once drilled, use a _____ with a diameter the same size as the rivet shank diameter to pry the head off, or tap the head lightly with a _____ to break it off from the shank.

(6) If the hole is damaged, use a _____ for the next larger diameter rivet to re-drill the hole. When using a larger diameter rivet, be sure that the _____, _____, and _____ values are all satisfactory for the pattern.

Ⅱ. Translate the following sentences into Chinese.

(1) Placing a straight edge on top of shop heads that have been driven in a row can check the uniformity of rivet heights. Each surface of the shop heads should touch the straightedge without gaps.

(2) When it has been determined that a rivet has been improperly installed, the rivet

must be removed and replaced without damaging the base metal.

(3) Use a drill the same diameter as the hole or one that is one number size smaller to drill down to the base of the rivet head.

(4) After the rivet head has been removed, back-up the underside of the skin with a bucking bar or piece of wood, and use a pin punch to gently drive the rivet shank from the sheet metal.

Lesson 12   Sheet Metal Repairs

INTENSIVE READING

SPECIFIC REPAIR TYPES

Before discussing any type of a specific repair that could be made on an aircraft, remember that the methods, procedures, and materials mentioned in the following paragraphs are only typical and should not be used as the authority for the repair. When repairing a damaged component or part, consult the applicable section of the manufacturer's structural repair manual for the aircraft. Normally, a similar repair will be illustrated, and the types of material, rivets, and rivet spacing and the methods and procedures to be used will be listed. Any additional knowledge needed to make a repair will also be detailed.

If the necessary information is not found in the structural repair manual, attempt to find a similar repair or assembly installed by the manufacturer of the aircraft.

Smooth Skin Repair

Minor damage to the outside skin of an aircraft can be repaired by applying a patch to the inside of the damaged sheet. A filler plug must be installed in the hole made by the removal of the damaged skin area. It plugs the hole and forms a smooth outside surface necessary for aerodynamic smoothness of modern day aircraft.

The size and shape of the patch is determined in general by the number of rivets required in the repair. If not otherwise specified, calculate the required number of rivets by using the rivet formula. Make the patch plate of the same material as the original skin and of the same thickness or of the next greater thickness.

Elongated Octagonal Patch

Whenever possible, use an elongated octagonal patch for repairing the smooth skin. This type of patch provides a good concentration of rivets within the critical stress area, eliminates dangerous stress concentrations, and is very simple to lay out. This patch may vary in length according to the condition of the repair.

Follow the steps shown in the paper layout of this patch (Figure 12 – 1). First, draw the outline of the trimmed-out damage. Then, using a spacing of three to four diameters of the rivet to be used, draw lines running parallel to the line of stress. Locate the lines for perpendicular rows two and one-half rivet diameters from each side of the cutout, and space the remaining lines three-fourths of the rivet pitch apart.

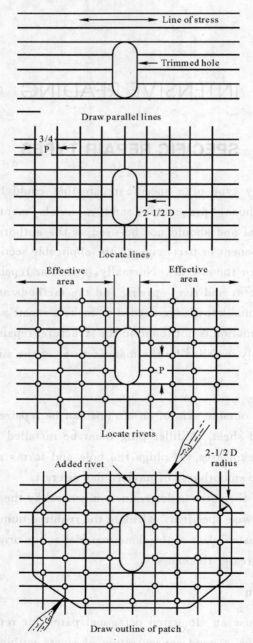

Figure 12 – 1　Elongated patch

Locate the rivet spots on alternate lines perpendicular to the stress lines to produce a stagger between the rows and to establish a distance between rivets (in the same row) of about six to eight rivet diameters. After locating the proper number of rivets on each side of the cutout, add a few more if necessary so that the rivet distribution will be uniform. At each of the eight corners, swing an arc of two and one-half rivet diameters from each corner rivet. This locates the edge of the patch. Using straight lines, connect these arcs to complete the layout.

Round Patch

Use the round patch for flush repairs of small holes in smooth sheet sections. The uniform distribution of rivets around its circumference makes it an ideal patch for places where the direction of the stress is unknown or where it is known to change frequently.

If a two-row round patch is used (Figure 12 - 2), first draw the outline of the trimmed area on paper. Draw two circles, one with a radius equal to the radius of the trimmed area plus the edge distance, and the other with a radius 3/4. in. larger. Determine the number of rivets to be used and space two-thirds of them equally along the outer row. Using any two adjacent rivet marks as centers, draw intersecting arcs; then draw a line from the point of intersection of the arcs to the center of the patch. Do the same with each of the other pairs of rivet marks. This will give half as many lines as there are rivets in the outer row. Locate rivets where these lines intersect the inner circle. Then transfer the layout to the patch material, adding regular outer edge material of two and one-half rivet diameters to the patch.

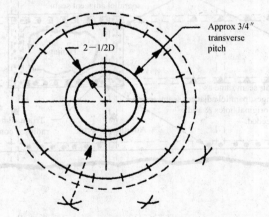

Figure 12 - 2   Layout of a two-row round patch

Panel Repair

In aircraft construction, a panel is any single sheet of metal covering. A panel section is the part of a panel between adjacent stringers and bulkheads. Where a section of skin is damaged to such an extent that it is impossible to install a standard skin repair, a special type of repair is necessary. The particular type of repair required depends on whether the damage is repairable outside the member, inside the member, or to the edges of the panel.

Damage which, after being trimmed, has less than eight and one-half manufacturer's rivet diameters of material inside the members requires a patch which extends over the members, plus an extra row of rivets along the outside of the members. For damage which, after being trimmed, has eight and one-half rivet diameters or more of material, extend the patch to include the manufacturer's row of rivets and add an extra row inside the member. Damage which extends to the edge of a panel requires only one row of rivets along the panel edge, unless the manufacturer used more than one row. The repair procedure for the other edges of the damage follows the previously explained methods.

The procedures for making all three types of panel repairs are similar. Trim out the damaged portion to the allowances mentioned in the preceding paragraph. For relief of stresses at the corners of the trim-out, round them to a minimum radius of 1/2 in. Lay out the new rivet row with a transverse pitch of approximately five rivet diameters and stagger the rivets with those put in by the manufacturer (Figure 12 - 3).

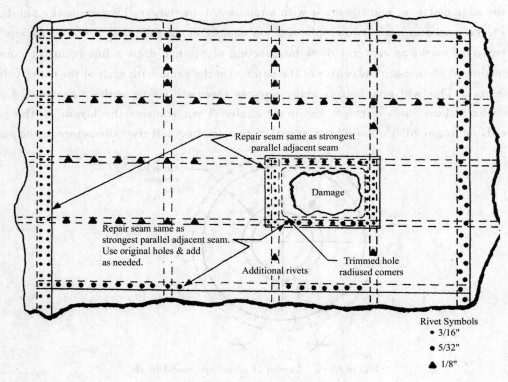

Repair seam same as strongest parallel adjacent seam

Damage

Repair seam same as strongest parallel adjacent seam. Use original holes & add as needed.

Additional rivets

Trimmed hole radiused corners

Rivet Symbols
• 3/16"
● 5/32"
▲ 1/8"

Figure 12 - 3 Panel skin patch

Spar Repair

The spar is the main supporting member of the wing. Other components may also have supporting members called spars which serve the same function as the spar does in the wing.

Think of spars as the "hub" or "base"of the section in which they are located, even though they are not in the center. The spar is usually the first member located during the construction of the section, and the other components are fastened directly or indirectly to it.

Because of the load the spar carries, it is very important that particular care be taken when repairing this member to ensure that the original strength of the structure is not impaired. The spar is so constructed that two general classes of repairs, web repairs and cap strip repairs, are usually necessary. An exploded view of a spar web butt splice is shown in figure 12－4. Figure 12－5 shows an exploded view of a T-spar cap strip repair.

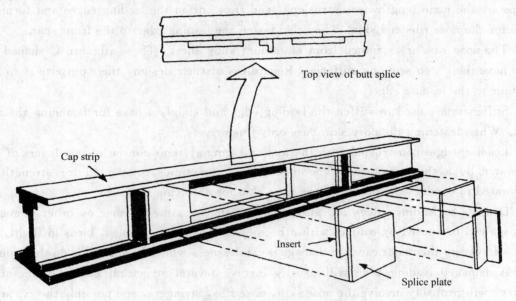

Figure 12－4   Spar web butt splice

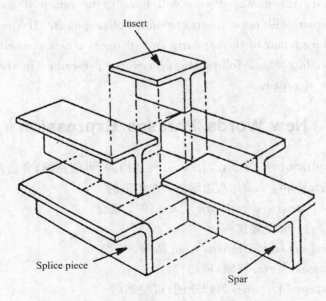

Figure 12－5   T-spar cap strip repair

Leading Edge Repair

The leading edge is the front section of a wing, stabilizer, or other airfoil. The purpose of the leading edge is to streamline the forward section of the wings or control surfaces so that the airflow is effective. The space within the leading edge is sometimes used to store fuel. This space may also house extra equipment such as landing lights, plumbing lines, or thermal anti-icing systems. The construction of the leading edge section varies with the type of aircraft. Generally, it will consist of cap strips, nose ribs, stringers, and skin. The cap strips are the main lengthwise extrusions, and they stiffen the leading edges and furnish a base for the nose ribs and skin. They also fasten the leading edge to the front spar.

The nose ribs are stamped from aluminum alloy sheet. These ribs are U-shaped and may have their web sections stiffened. Regardless of their design, their purpose is to give contour to the leading edge.

Stiffeners are used to stiffen the leading edge and supply a base for fastening the nose skin. When fastening the nose skin, use only flush rivets.

Leading edges constructed with thermal anti-icing systems consist of two layers of skin separated by a thin air space. The inner skin, sometimes corrugated for strength, is perforated to conduct the hot air to the nose skin for anti-icing purposes.

Damage to leading edges are also classified in the same manner as other damages. Damage can be caused by contact with other objects, namely, pebbles, birds in flight, and hail. However, the major cause of damage is carelessness while the aircraft is on the ground.

A damaged leading edge will usually involve several structural parts. Flying-object damage will probably involve the nose skin, nose ribs, stringers, and possibly the cap strip. Damage involving all of these members will necessitate installing an access door to make the repair possible. First, the damaged area will have to be removed and repair procedures established. The repair will need insertions and splice pieces. If the damage is serious enough, it may require repair of the cap strip and stringer, a new nose rib, and a skin panel. When repairing a leading edge, follow the procedures prescribed in the appropriate repair manual for this type of repair.

New Words/Phrases/Expression

1. applicable[ə'plɪkəb(ə)l; 'æplɪk-]   *adj.* 可适用的；可应用的；合适的
2. octagonal[ɔk'tægənl]   *adj.* 八边形的，八角形的
3. concentration[kɒns(ə)n'treɪʃ(ə)n]   *n.* 集中，集合
4. stagger['stægə]   *n.* 交错排列
5. circumference[sə'kʌmf(ə)r(ə)ns]   *n.* 圆周，周围
6. intersect[ɪntə'sekt]   *vt.* 横断，横切；贯穿
7. repairable[ri'pɛərəbl]   *adj.* 可修理的；可挽回的

8. hub[hʌb]   *n.* 中心；毂；木片

9. butt[bʌt]   *n.* 粗大的一端，靶垛

10. pattern['pæt(ə)n]   *n.* 模式；图案；样品

11. extrusion[ɪk'stru:ʒn]   *n.* 挤出；推出；赶出；喷出

12. intrusion[ɪn'tru:ʒ(ə)n]   *n.* 闯入，侵扰

13. stiffen['stɪf(ə)n]   *vt.* 使变硬；使黏稠

14. corrugate['kɒrugeɪt]   *vt.* 使起皱；成波状

15. perforate['pɜːfəreɪt]   *vt.* 穿孔于，打孔穿透

16. conduct['kɒndʌkt]   *vt.* 管理；引导；表现

17. pebbles['peb(ə)l]   *n.* 卵石

18. hail[heɪl]   *n.* 冰雹

Notes

（1）Before discussing any type of a specific repair that could be made on an aircraft, remember that the methods, procedures, and materials mentioned in the following paragraphs are only typical and should not be used as the authority for the repair.

分析：第一个 that 为定语从句；第二个 that 为宾语从句。

翻译：在讨论飞机上可以做的任何类型的特定修理之前，请记住，在下面的段落中提到的方法，程序和材料仅仅是典型的，不应该被用来作为维修的权威。

（2）A filler plug must be installed in the hole made by the removal of the damaged skin area.

分析：made by 为定语。

翻译：在被损伤蒙皮区域去除后所形成的洞里，必须安装一个填充塞。

（3）Whenever possible, use an elongated octagonal patch for repairing the smooth skin. This type of patch provides a good concentration of rivets within the critical stress area, eliminates dangerous stress concentrations, and is very simple to lay out.

分析：Whenever possible 只要有可能。

翻译：只要有可能，尽量使用一个加长的八角形补片来修理光滑的蒙皮。这种类型的补片为严重应力区域内的铆钉提供了一个好的集中效果，消除了危险的应力集中，是非常简单的布局。

（4）Because of the load the spar carries, it is very important that particular care be taken when repairing this member to ensure that the original strength of the structure is not impaired.

分析：the spar carries 是 the load 的定语。

翻译：因为翼梁传递的载荷，所以当修理这个构件时要特别小心，确保这个结构的原有强度不受损，这是非常重要的。

Exercise

Ⅰ. Answer the following questions.

(1)Why must a filler plug be installed in the hole made by the removal of the damaged skin area? How to determine the size and shape of the patch?

(2)What's the advantage of the elongated octagonal patch?

(3)What's the advantage of the round patch?

(4)How many types of panel repairs are there? What are the procedures?

(5)What's the purpose of the leading edge? What is the space within the leading edge used for?

Ⅱ. Put the following sentences into Chinese.

(1)When repairing a damaged component or part, consult the applicable section of the manufacturer's structural repair manual for the aircraft.

(2)At each of the eight corners, swing an arc of two and one-half rivet diameters from each corner rivet.

(3)Because of the load the spar carries, it is very important that particular care be taken when repairing this member to ensure that the original strength of the structure is not impaired.

(4)Damage can be caused by contact with other objects, namely, pebbles, birds in flight, and hail. However, the major cause of damage is carelessness while the aircraft is on the ground.

EXTENSIVE READING

BASIC PRINCIPLES OF SHEET METAL REPAIR

Methods of repairing structural portions of an aircraft are numerous and varied, and no set of specific repair patterns has been found which will apply in all cases. Since design loads acting in various structural parts of an aircraft are not always available, the problem of repairing a damaged section must usually be solved by duplicating the original part in strength, kind of material, and dimensions. Some general rules concerning the selection of material and the forming of parts which may be applied universally by the airframe mechanic will be considered in this chapter.

Basic Principles of Sheet Metal Repair

The repairs discussed are typical of those used in aircraft maintenance and are included to introduce some of the operations involved. For exact information about specific repairs, consult the manufacturer's maintenance or service manuals.

The first and one of the most important steps in repairing structural damage is "sizing up" the job and making an accurate estimate of what is to be done. This sizing up includes an estimate of the best type and shape of patch to use; the type, size, and number of rivets needed; and the strength, thickness, and kind of material required to make the repaired member no heavier (or only slightly heavier) and just as strong as the original. Also inspect the surrounding members for evidence of corrosion and load damage so that the required extent of the "clean-out" of the old damage can be estimated accurately. After completing the clean-out, first make the layout of the patch on paper, then transfer it to the sheet stock selected. Then, cut and chamfer the patch, form it so that it matches the contour of that particular area, and apply it.

Maintaining Original Strength

In making any repair, certain fundamental rules must be observed if the original strength of the structure is to be maintained. The patch plate should have a cross-sectional area equal to, or greater than, that of the original damaged section. If the member is subjected to compression or to bending loads, place the splice on the outside of the member to secure a higher resistance to such loads. If the splice cannot be placed on the outside of the member, use material that is stronger than the material used in the original member.

To reduce the possibility of cracks starting from the corners of cutouts, try to make cutouts either circular or oval in shape. Where it is necessary to use a rectangular cutout, make the radius of curvature at each corner no smaller than 1/2 in. Either replace buckled or bent members or reinforce them by attaching a splice over the affected area.

Be sure the material used in all replacements or reinforcements is similar to the material used in the original structure. If it is necessary to substitute an alloy weaker than the original, use material of a heavier gage to give equivalent cross-sectional strength. But never practice the reverse; that is, never substitute a lighter gage stronger material for the original, this apparent inconsistency is because one material can have greater tensile strength than another, but less compressive strength , or vice versa. As an example, the mechanical properties of alloys 2024 – T and 2024 – T80 are compared in the following paragraph.

If alloy 2024 – T were substituted for alloy 2024 – T80, the substituted material would have to be thicker unless the reduction in compressive strength was known to be acceptable. On the other hand, if 2024 – T80 material were substituted for 2024 – T stock, the substituted material would have to be thicker unless the reduction in tensile strength was known to be acceptable. Similarly, the buckling and torsional strength of many sheet-metal and tubular parts are dependent primarily upon the thickness rather than the allowable compressive and shear strengths.

When forming is necessary, be particularly careful, for heat-treated and cold-worked alloys will stand very little bending without cracking. Soft alloys, on the other hand, are easily formed but are not strong enough for primary structures. Strong alloys can be formed in their annealed condition and heat-treated to develop their strength before assembling.

In some cases, if the annealed metal is not available, heat the metal, quench it according to regular heat-treating practices, and form itbefore age-hardening sets in. The forming should be completed in about half an hour after quenching, or the material will become too hard to work.

Maintaining Original Contour

Form all repairs in such a manner that they will fit the original contour perfectly. A smooth contour is especially desirable when making patches on the smooth external skin of high-speed aircraft.

Keeping Weight to a Minimum

Keep the weight of all repairs to a minimum. Make the size of the patches as small as practicable and use no more rivets than are necessary. In many cases, repairs disturb the original balance of the structure. The addition of excessive weight in each repair may unbalance the aircraft so much that it will require adjustment of the trim-and-balance tabs. In areas such as the spinner on the propeller, a repair will require application of balancing patches so that a perfect balance of the propeller assembly can be maintained.

General Structural Repair

Aircraft structure members are designed to perform a specific function or to serve a definite purpose. The prime objective of aircraft repair is to restore damaged parts to their original condition. Very often, replacement is the only way in which this can be done effectively. When repair of a damaged part is possible, first study the part carefully so that its purpose or function is fully understood.

Strength may be the principal requirement in the repair of certain structures, while others may need entirely different qualities. For example, fuel tanks and floats must be protected against leakage; but cowlings, fairings, and similar parts must have such properties as neat appearance, streamlined shape, and accessibility. The function of any damaged part must be carefully determined so that the repair will meet the requirement

New Words/Phrases/Expression

1. numerous['nju:m(ə)rəs] *adj.* 许多的，无数的
2. portion['pɔːʃ(ə)n]   *n.* 一部分，一份
3. duplicate['djuːplɪkeɪt]   *vt.* 复写，复制，使加倍
4. accurate['ækjʊrət] *adj.* 正确的,精确的

5. clean-out　清除,扫荡

6. chamfer['tʃæmfə]　*vt.* 去角;挖槽;斜切

7. contour['kɒntʊə]　*n.* 轮廓,周线,等高线

8. cutout['kʌt,aʊt]　*n.* 挖去部分;删除部分;开孔

9. rectangular[rek'tæŋgjʊlə]　*adj.* 矩形的,成直角的

10. curvature['kɜːvətʃə]　*n.* 弯曲,曲率

11. layout['leɪaʊt]　*n.* 布局;设计;安排;陈列

12. reinforcement[riːɪn'fɔːsm(ə)nt]　*n.* 加固;增援;援军;加强

13. inconsistency[ɪnkən'sɪst(ə)nsɪ]　*n.* 不一致;易变

14. substitute['sʌbstɪtjuːt]　*vt.* 代替

15. buckle['bʌk(ə)l]　*vt.* 扣住;使弯曲

16. definite['defɪnɪt]　*adj.* 明确的,一定的

17. spinner['spɪnə]　*n.* 螺旋桨整流罩,桨毂盖

Exercises

Ⅰ. Answer the following questions.

(1) What are the basic principles of sheet metal repair?

(2) What is the first and one of the most important steps in repairing structural damage?

(3) What would be done if alloy 2024 – T were substituted for alloy 2024 – T80?

(4) What would be done if 2024 – T80 material were substituted for 2024 – T?

(5) What shall we do if the annealed metal is not available?

Ⅱ. Put the following sentences into Chinese.

(1) Methods of repairing structural portions of an aircraft are numerous and varied, and no set of specific repair patterns has been found which will apply in all cases.

(2) Where it is necessary to use a rectangular cutout, make the radius of curvature at each corner no smaller than 1/2 in.

(3) The addition of excessive weight in each repair may unbalance the aircraft so much that it will require adjustment of the trim-and-balance tabs.

(4) Form all repairs in such a manner that they will fit the original contour perfectly. A smooth contour is especially desirable when making patches on the smooth external skin of high-speed aircraft.

Lesson 13   Special Aircraft Repair Tools

INTENSIVE READING

SPECIAL AIRCRAFT REPAIR TOOLS

Deburring Tool

If burrs, slivers which form at the edges of holes whenever a drill breaks through the metal, are not removed, they will deform the joint by creating a bulge between the skins when the fasteners are installed into the holes. A good deburring tool is a twist drill one or two time larger than the hole. Insert it into the hole and spin by hand until all the burrs are removed (Figure 13 – 1).

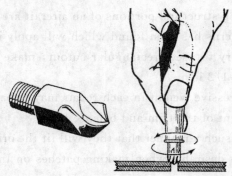

Figure 13 – 1   Deburring tool

Microstop Countersink Cutters

There are three types of countersink cutters: a non-adjustable fixed cutter, a stop countersink which is equipped with a rivet hole pilot independent of the cutter, and a microstop countersink cutter which includes a hole pilot. The mictostop countersink cutter

(Figure 13 – 2) allows the user to make micro adjustments when setting the depth of the countersink cutter. Another advantage of the microstop countersink is that it is more durable than the other types.

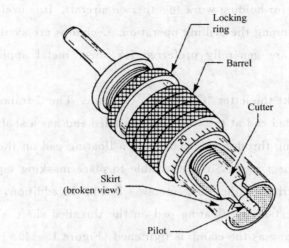

Figure 13 – 2   Microstop countersink

Cleco Clampsand Pliers

Cleco clamps are used to align parts prior to being reriveted to an aircraft. The clamps are installed with Cleco pliers (Figure 13 – 3). The color of the Cleco clamp indicates the diameter of the rivet it is to be used with. Four commonly used sizes are 3/32 of an inch (silver), 1/8 of an inch (copper), 5/32 of an inch (black), and 3/16 of an inch (brass, gold).

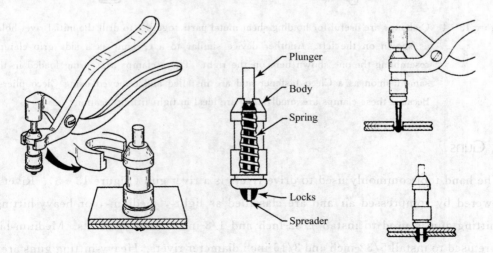

Figure 13 – 3   Cleco fasteners are used to temporarily hold sheet metal parts together until they are riveted

C-Clamps

The C-clamp is a tool primarily used by machinists, but has been adapted by technicians working with sheet metal for holding work together on aircraft. It is useful for holding sheet metal in place before beginning the drilling operation. C-clamps are available in many sizes. However, smaller sizes are generally preferred for sheet metal applications to prevent damage to the metal.

The C-clamp looks like the letter "C"; hence, its name. The C-frame has a fixed rest on its lower end and a threaded end at the top. The threaded end has a shaft that runs through it with a tee handle running through the shaft, and a floating pad on the end. Before using one of these clamps on sheet metal, it is advisable to place masking tape over each of the pads to help prevent marring of the sheet metal's finish. In addition, before using these clamps, check to make certain the floating pad on the threaded shaft is free to swivel and turn to help prevent marring as the clamp is tightened (Figure 13 – 4).

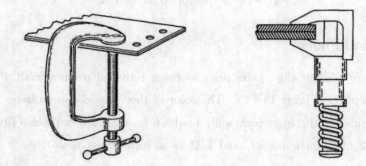

Figure 13 – 4　C-clamps are useful for holding sheet metal parts together to drill the initial rivet holes as shown on the left. Another device similar to a C-clamp is a side grip clamp, resembling the one shown here on the right. These clamps are spring loaded in the same fashion as a Cleco fastener and are installed and removed using Cleco pliers. Because these clamps are small, they are ideal in tight fitting locations

Rivet Guns

The hand tool commonly used to drive rivets is a rivet gun (Figure 13 – 5). Rivet guns are powered by compressed air and are classified as light-, medium-, or heavy-hitting. A light-hitting gun is used to install 1/32-inch and 1/8-inch diameter rivets. Medium-hitting guns are used to install 5/32-inch and 3/16-inch diameter rivets. Heavy-hitting guns are used to install larger diameter rivets and some special fasteners.

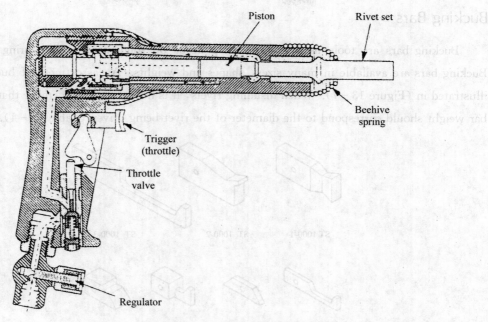

Figure 13 – 5   Rivet gun

Rivet Cutters

Rivet cutters are used to cut rivets to size prior to driving. The rivet cutter has a stack of thickness gauges which are used to determine the correct rivet length by measuring the space between the rivet head and the cutting edge (Figure 13 – 6).

When rivet cutters are not available, the rivets can be cut to size using a pair of diagonal cutting pliers. The rivet is cut by squeezing together the two rotating plates connected to the cutter handles.

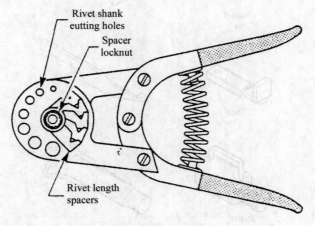

Figure 13 – 6   Rivet cutters

Bucking Bars

Bucking bars are tools used to form shop heads on solid-shank rivets during installation. Bucking bars are available in many sizes, shapes and weights. An assortment of bucking bars is illustrated in (Figure 13 – 7). When installing rivets, it is important to remember that the bucking bar weight should correspond to the diameter of the rivet being driven (Table 13 – 1).

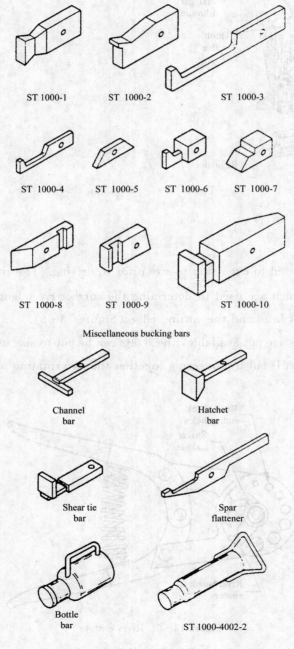

ST 1000-1   ST 1000-2   ST 1000-3

ST 1000-4   ST 1000-5   ST 1000-6   ST 1000-7

ST 1000-8   ST 1000-9   ST 1000-10

Miscellaneous bucking bars

Channel bar

Hatchet bar

Shear tie bar

Spar flattener

Bottle bar

ST 1000-4002-2

Figure 13 – 7   Bucking bars

Table 13 - 1   Rivet information chavt

Diameter of rivet	Rivet diameter decimal	Drill decimal	Oversize of drill	Drill number	Cleco color	Bucking bar weight	Minimum rivet pitch	Minimum edge dist
3/32	.0937	.0980	.0043	40	SILVER	2—3 lbs.	9.32	6/32
1/8	.125	.1285	.0035	30	COPPER	3—4 lbs.	3/8	2/8
5/32	.15625	.1590	.0028	21	BLACK	3—4 lbs.	15/32	10/32
3/16	.1875	.1910	.0035	11	GOLD	4—5 lbs.	9/16	6/16
1/4	.250	.2570	.007	F	GREEN	5—6.5 lbs.	3/4	2/4

RIVET—DRILL—CLECO—BUCKING BARS—MINIMUMS

Note: alcoa claims that a rivet should fit as tight as possible before driving, especially those rivets of harder alloy. the hole clearances listed above are the recommended sizes used in the field.

The face of the bucking bar is smooth and must be protected from any nicks or scars which can affect the proper shaping of the driven head. Damaged bucking bar faces can be filed and then smoothed using oil and crocus cloth.

Nibblers

Nibblers, as the name suggests cut metal by nibbling away small pieces until the correct size is obtained. A typical hand nibbler (Figure 13 - 8) is operated by squeezing repeated notches out of the metal. Nibblers can be used to enlarge an inspection hole or to cut a notch to fit around an obstruction.

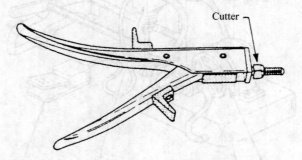

Cutter

Figure 13 - 8   Hand nibblers

Bench And Floor Tools

Squaring Shears

Squaring shears can be operated either manually or hydraulically. A foot-operated treadle can be inexpensive converted to a hydraulic type by installing a hydraulic power pack on the machine. A hydraulic squaring shear can cut aluminum alloy up to 1/8 of an inch thick and aircraft steel up to 16 gauge. Figure 13 – 9 shows a squaring shear. The primary purpose of the side fence is to square metal to the cutting blade. A piece of sheet metal can be squared by placing it beyond the cutter, shearing off one side, and turning the metal 90 degree to make another cut.

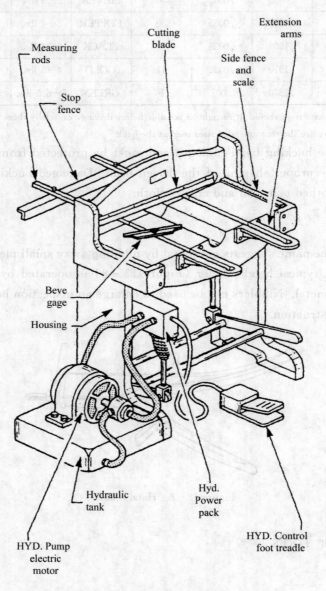

Figure 13 – 9   Squaring shear

Extending from the back of the machine are two rods, which, like the side fence, have a measuring scale embossed on them. Connected to these two rods is a stop fence that can be adjusted for longer cuts. The rods can be set to measure the distance from the cutting blade to the adjusted stop.

For example, if a sheet metal skin is to be cut 10 inches long and has to be squared, slide the skin along the fixed side fence, pushing it until it contacts the stop fence, which has been set at 10 inches. Then cut by stepping in the treadle.

Hole Finder Or Duplicator

Hole finders, or duplicators, are used to locate holes to be made in undrilled skin. Hole finders can be easily and inexpensively made to meet specific needs. Figure13 − 10(a) shows a typical homemade hole finder. The holes in the rib or stringer to be reused serve as the alignment holes. The duplicator marks the location of the holes to be made in the new skin. To use a hole duplicator, first insert the rivet alignment pin into one of the holes on the rib. Then place the undrilled skin over the hole finder's rivet head. Finally, center-punch the

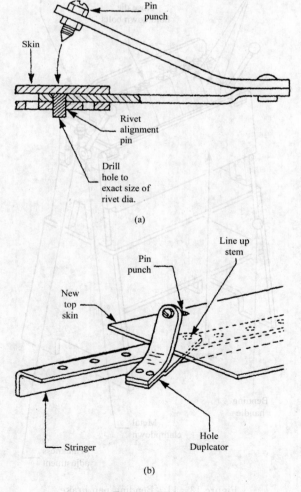

Figure   13 − 10

(a) Hole finder; (b) Hole finder used to duplicate holes

new skin (Figure 13 – 10(b)).

Bending Brakes

There are many different sizes and types of bending brakes. The most versatile and widely used is the box and pan brake (Figure 13 – 11). The box and pan brake is fully adjustable and comes in various lengths. It has removable nose pieces that make the construction of a pan rib possible. The whole brake section can be moved backwards and forwards, allowing the addition of thicknesses of metal to the nose of the brake to increase the radius of the bend (Figure 13 – 12). As the bend radius is increased, the nose brake clearance must also be increased.

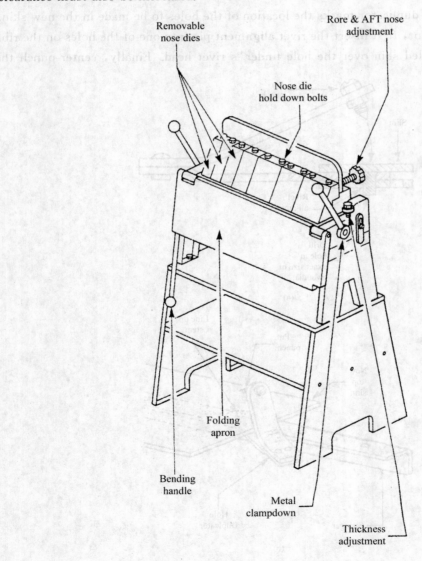

Figure 13 – 11 Bending pan brake

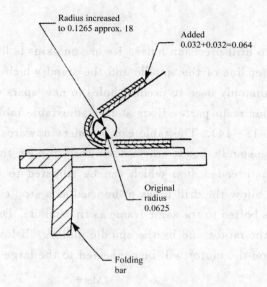

Radius increased
to 0.1265 approx. 18

Added
0.032+0.032=0.064

Original
radius
0.0625

Folding
bar

Figure13 – 12　Increasing bend radius bar

Straight bending brakes range from four to twelve feet in length. They are usually not adjustable, and it is difficult to install a bend radius bar on the nose of the brake. The straight bending brake is used to make long stringers.

Slip Roller

Slip roller (Figure 13 – 13) is used to make a cylinder or a duct. The slip roller consists of three rollers interconnected by gears and turned by a hand crank. The rollers can be individually adjusted to produce a tight or a loose roll. After a metal cylinder is formed, it is removed from the roller by activating a release which allows the roller to slip free of its holder at one end. The part can then be removed.

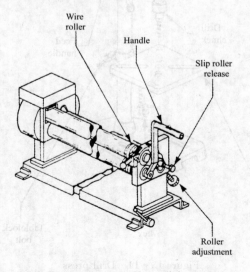

Wire
roller

Handle

Slip roller
release

Roller
adjustment

Figure 13 – 13　Slip roll former

Drill Press

A drill press is used to drill precision holes. Its use on skins is limited by the reach, or distance, between the center line of the spindle and the stand which holds the drill's drive mechanism. It is more commonly used to predrill holes in new spars or stringers.

The drill press has four main parts: floor stand, adjustable table, adjustable spindle, and electric motor (Figure 13 – 14). The table can be raises upwards toward the spindle to reduce the distance that the spindle has to travel downward to meet the work. On some drill presses, the spindle has a threaded stop which can be adjusted to control the amount of downward travel and thus allow the drill press to be used as a stop countersink cutter. The drill press electric motor is bolted to the same frame as the spindle. Drill speed is determined by the belt's location on the motor and by the spindle pulleys. If low speed is required for drilling, the small pulley on the motor will be connected to the large pulley on the spindle.

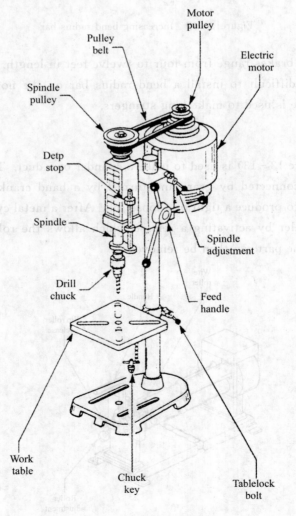

Figure 13 – 14　Drill press

Combination Band Saw

Acombination hack and band saw (Figure 13 – 15) is used when thick sheet metal cannot be cut with the squaring shears. A medium-size combination saw will do most such cutting jobs.

Electric Grinders

The electric grinder is used regularly to sharpen dull or broken drills. A grinder does not have to be large; a small 1/2 to 3/4 horsepower motor is adequate. The grinding wheel best suited for sharpening drills is a fine grit stone. When a grinding wheel becomes grooved and uneven, it should be re-dressed and squared with a dressing tool. Care should be taken to avoid dropping the grinding wheel because it can develop small cracks which could later cause the wheel to fly apart and seriously injure the operator.

Note: never use a grinding wheel that has been dropped or that has a flaw of any kind.

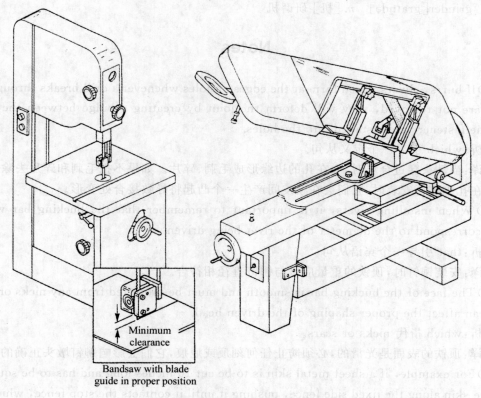

Minimum clearance

Bandsaw with blade guide in proper position

Figure 13 – 15　Band saw

New Words/Phrases/Expression

1. webbing['webɪŋ]   *n.* 带子;边带;厚边
2. clamp[klæmp]   *n.* 夹钳;螺丝钳;夹具
3. chuck[tʃʌk]   *n.* [机]卡盘
4. plexiglass['pleksɪɡlɑːs, -ɡlæs]   *n.* 树脂(有机)玻璃
5. deburr[diːˈbɜː]   *vt.* 去毛刺;倒角;清理毛刺
6. countersink['kaʊntəsɪŋk]   *n.* 埋头孔
7. assortment[əˈsɔːtm(ə)nt]   *n.* 分类;混合物
8. nibbler['nɪblə]   *n.* 切片机;[机]毛坯下料机
9. notch[nɒtʃ]   *n.* 刻痕;凹口
10. hydraulic[haɪˈdrɒlɪk]   *adj.* 液压的
11. pulley['pʊlɪ]   *n.* 滑轮;皮带轮
12. grinder['ɡraɪndə]   *n.* [机]研磨机

Notes

(1)If burrs, slivers which form at the edges of holes whenever a drill breaks through the metal, are not removed, they will deform the joint by creating a bulge between the skins when the fasteners are installed into the holes.

分析:which 引导一个定义从句。

翻译:当钻头穿过金属时,会在孔的边缘形成毛刺、碎片。如果不把毛刺和碎片去除掉,当紧固件在孔中进行安装时,它们会在蒙皮间产生一个凸起将导致接合处变形。

(2)When installing rivets, it is important to remember that the bucking bar weight should correspond to the diameter of the rivet being driven.

分析:that 引导一个宾语从句。

翻译:安装铆钉时,顶铁的重量应该与铆钉直径相符合。

(3)The face of the bucking bar is smooth and must be protected from any nicks or scars which can affect the proper shaping of the driven head.

分析:which 指代 nicks or scars。

翻译:顶铁的表面是光滑的,必须防止任何刻痕或疤痕,它们会影响铆钉墩头正确的形状。

(4)For example, if a sheet metal skin is to be cut 10 inches long and has to be squared, slide the skin along the fixed side fence, pushing it until it contacts the stop fence, which has been set at 10 inches. Then cut by stepping in the treadle.

分析:which 引导一个非限制性定语从句。

翻译:例如,如果一个薄金属板蒙皮要被切割成 10 英寸长,并且必须使各边成直角,可沿固定的侧边围栏滑动蒙皮,推动它,直到它与止动围栏接触。该止动围栏已被设置为 10 英寸。然后踏动踏板进行切割。

Exercises

I. Answer the following questions.

(1) What are hole finders used for?

(2) Which diameter rivets is a light-hitting gun used to install?

(3) How to deal with the damaged bucking bar faces?

(4) What's the purpose of the side fence of squaring shears?

II. Put the following sentences into Chinese.

(1) In aircraft repair work, the air-driven motor is usually preferred because electrical lines can be accidentally cut while working in close quarters on an aircraft, and because all air motors have variable drill speeds and are more durable than most electrical motors.

(2) There are three types of countersink cutters: a non-adjustable fixed cutter, a stop countersink which is equipped with a rivet hole pilot independent of the cutter, and a microstop countersink cutter which includes a hole pilot.

(3) Medium-hitting guns are used to install 5/32 – inch and 3/16 – inch diameter rivets. Heavy-hitting guns are used to install larger diameter rivets and some special fasteners.

(4) To protect the teeth, files should be stored separately in a plastic wrap or hung by their handles. Files which are kept in a toolbox should be wrapped in waxed paper to prevent rust from forming on the teeth.

(5) Drill speed is determined by the belt's location on the motor and by the spindle pulleys. If low speed is required for drilling, the small pulley on the motor will be connected to the large pulley on the spindle.

(6) When a braced wing needs to be repaired and the strut must be removed, a wing jig or stand is placed underneath one of the compression ribs.

EXTENSIVE READING

Just How Safe is Air Travel?

With all the publicity, press attention and public enquiries that accompany every air accident, and with all that is written and said about emergencies, it is easy to become obsessed by safety precautions, and to overlook the fact that per passenger mile, air travel is much safer than traveling by road. The chances of an accident happening to an aircraft are relatively small.

But accidents do occur, and attract considerable publicity. The airline concerned may lose a valuable aircraft and an even more valuable aircrew; it will also suffer from what is called the 'follow on' effect, which, indirectly, affects other airlines too. Some people, after reading or hearing the news of a crash, just will not fly, and this is unreasonable in view of the statistics.

Improved aircraft design and improved maintenance have gone a long way to reducing crashes and the fatalities caused by them. Today, less than two percent of all fatal crashes are attributable to faulty maintenance. New flying techniques can deal with any unpredictable weather conditions, and the introduction of ground-based computers in some countries means that aircraft performance can be monitored in flight and potential malfunctions can be forecast and appropriate action taken to rectify them.

Technology has made aviation today one of the safest modes of travel. However, it has to be admitted that between sixty and eighty percent of all air crashes are due to human error. Crew training is now concentrating on this area as the extensive use of flight simulators demonstrates.

Flight simulators, although expensive to install, have cut down the overall costs of training on one hand and, on the other, improved the quality of the training by including emergency simulations which would be impossible to carry out in real life situations. This simulated training includes handling the aircraft under severe weather conditions such as ice accretion, wind shear and clear air turbulence, or dealing with any malfunction of any of the aircraft's systems or powerplant. If this training is repeated regularly, and updated when the introduction of more modern techniques requires it, flight crews will be ready to deal with many emergency situations which are potentially dangerous, even fatal.

Some accidents are, moreover, caused by situations where the crew—however well trained individually—fail to act as a team. People often react quite differently when under stress than when under normal conditions. Anxiety about the health of a member of his family may cause a pilot to lose attention and commit an error he would not otherwise have made. Tensions due to personality clashes between crew members may lead to misunderstandings at critical moments and an accident, which could otherwise have been avoided, happens. A simulator can help to identify the areas of possible stress, and also to identify crew members who are prone to, or susceptible to showing weaknesses in stressful situations.

But flight crews are not the only people who need to train for emergencies. Good cabin crew training will go a long way to reducing the numbers of fatalities occurring in crashes which many may be fortunate to survive. And surviving a crash will depend very much on the cabin crew's ability to assess the nature of the emergency, to react swiftly and with presence of mind, and to make the proper use of the emergency equipment. As for flight crews, it is impossible to assess a cabin crew's behavior under stress unless emergency situations can be simulated and crews exposed to them.

There is today an increasing number of cabin emergency simulators on the market, though not as many as there are flight deck simulators. But whereas the latter must reproduce all the controls, instruments and flight characteristics of the particular aircraft for which the crew are being trained, cabin simulators can be more standardized, require little complex instrumentation and no flight controls, and are therefore cheaper, though a completely equipped cabin simulator is still a very expensive item. To be effective, they must be able to simulate as accurately as possible all the conditions that might arise in any in-flight emergency - even to the extent of a fire in a toilet. In addition to all the normal cabin fitments, galleys, toilets, seating and equipment, full emergency equipment must also be carried-life vests, rafts, evacuation chutes, etc. Simulated flight will include the real sensations of taxiing, take-off, acceleration, climb, various degrees of turbulence, descent, approach, landing bump and deceleration during roll-out. During this simulated flight, the cabin crew can be trained to carry out their normal duties in the proper sequence. An otherwise normal flight can be made to end in a simulated crash (such as caused by landing gear failure, or ditching at sea) with rapid deceleration, crash noises, smoke and even fire. (Noise and lighting are used to simulate fire). Real—but not toxic—smoke can be released from canisters obliging the cabin crew to put on goggles and masks, often in total darkness, find and operate emergency doors and give instructions to passengers about the evacuation of the aircraft, use of chutes, etc. Injured passengers, also simulated, may have to be dealt with. Some training institutions include a swimming pool as part of the facilities in order to familiarize cabin staff with ditching procedure.

It will be seen from all this that most airlines take the training of their crews very seriously, almost regardless of initial costs. In the long run, the reputation that an airline has — not only for safety, but also for the attention it gives to emergency training —will win passengers from the competition in the same way that good cabin service will. No modern airline can afford to ignore this fact. Moreover this training, which is still being perfected, will go a long way to ensuring that aviation is the safest form of travel, and will remain so.

Exercises

Answer the following according to the instructions given after each.

(1) (a) Disasters involving aircraft attract a lot of publicity which hides the fact that air travel is much safer than people think it is.

(b) Because accidents to aircraft attract a lot of publicity, they are also thought to be very dangerous.

(c) People become obsessed with the dangers of traveling by air but the same people don't worry about traveling by road.

(Tick one)

(2) (a) At best, air travel can be said to be no more than a comparatively safe form of transport.

(b) There are fewer air accidents in one year than there are road accidents, which is why air travel is much safer.

(c) Statistics comparing air with road travel point to the fact that air travel is safer.

(Tick one)

(3) (a) Losing passengers after an air crash is much worse for an airline than losing a crew.

(b) The long-term effects of an air accident can be more serious for an airline than its immediate effects.

(c) The follow-on effect after an accident involving one airline are more serious for other airlines.

(Tick one)

(4) (a) There are always people who are stupid enough to listen to news about air accidents and refuse to fly.

(b) Only some unreasonable people are stupid enough not to want to fly as the result of reading about an air accident.

(c) Some people are unreasonable in refusing to fly as the result of having read about an air accident.

(Tick one)

(5) The public can over-react to excessive publicity about an accident involving aircraft and passenger fatalities, and this may affect air travel adversely.

(True/ False)

(6) (a) Errors due to poor maintenance account for less than 2 percent of all accidents involving deaths in an air crash.

(b) Deaths from crashes caused by faulty maintenance represent no more than 2 percent of the whole.

(c) No more than 2 percent of all crashes are fatal and these are directly attributable to faulty maintenance.

(Tick one)

(7) (a) Uncertain weather conditions can be handled by modern flying techniques.

(b) New flying techniques are such that all weather conditions can be predicted.

(c) However uncertain the weather, there are new ways of flying the aircraft so that the effects can be avoided.

(Tick one)

(8) It is possible nowadays to keep a computer check from the ground on all flights and so foresee the possibility of an emergency arising, and cause measures to be taken to prevent an accident

(True/ False)

(9) (a) Something around three quarters of ail air crashes are the result of mistakes attributable to men.

(b) Between 60 and 80 percent of all human errors lead to an air crash.

(c) Too many air crashes are caused by people's careless mistakes.

(Tick one)

Lesson 14   Composite Repairs

INTENSIVE READING

Composite Repairs

Introduction

With the increasing use of composite structures in aircraft components, it has become necessary to develop repair methods that will restore the component's original design strength without compromising its structural integrity. One of the main concerns is whether large repairs are always necessary to restore strength or whether smaller, less intrusive repairs can be implemented instead. With the concerns in mind, the main objective of this chapter was to discuss the steps of repairs applied to composite structures. The damage evaluation process (DEP) is a step-by-step procedure to make an analysis of the damage to structure and find the correct action. The overall contents were as follows: Types of Damage Identification, Damage Detection, Component Identification, Paint Removal, Damage Removal, Repair Design, Vacuum Bagging Materials for Composite Repair, Repair Instructions, Scarfing, the Repair Patch and Curing Methods.

Types of Damage

Chapter 5 of the maintenance manual shows inspections for damage incidents that can cause hidden damage that is more than the structure was first designed for. Most composites are damaged while in-service from overloading, impact or environmental effects. Parts may also need to be repaired due to a manufacturing defect. Such defects may make a composite structure more susceptible to in-service damage.

Holes and Punctures

Holes and punctures are usually caused by high or medium level impacts. They can be severe, but are usually easily detected.

Delaminations

Perhaps the most common of all damage types, delaminations are often caused by low-energy impacts, such as a tool drop or a glancing bird strike on an aircraft. They can sometimes be visible if they are near the surface. Delamination can also be caused by manufacturing defects, resin starvation in a laminate and by moisture ingression. (See Figure 14 – 1. Top left and right — this aramid-skinned/foam-cored panel illustrates how the plies of a laminate can actually come apart, or de-laminate. Bottom — carbon fiber sample shows how a localized impact may cause delamination within the laminate across a significant area.)

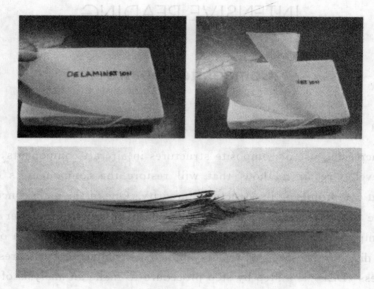

Figure 14 – 1   Delamination damage

Disbonds

A disbond indicates an adhesive bond failure between dissimilar materials. This is usually seen as a face sheet disbonding from an underlying sandwich core material. (See Figure 14 – 2. Top — a face skin disbonding from foam core. Bottom — Close-up showing that the core was once adhered to the skin.)

Core Damage

Core damage can occur with any type of core, and is typically caused by improper vacuum bagging, handling damage in manufacturing, and impacts in-service. Fluid ingress is a common problem with honeycomb core panels. This can facilitate extended damage, from freeze-thaw cycling in aircraft parts (Figure 14 – 3).

Figure 14 − 2   Disbond damage

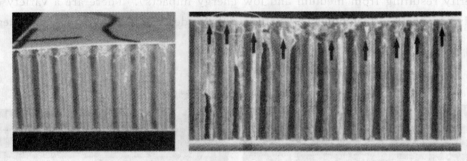

Figure 14 − 3   Core damage to honeycomb core caused by impact

Resin Damage

Resin damage is caused by many factors, such as fire or excessive heat, UV rays, paint stripper, or impacts. This type of damage may be hard to detect, and it is especially difficult to quantify its effects on the structural integrity of the part. As a general rule, resin damage leads to a greater loss in compressive strength than in tensile strength.

Water Ingression or Intrusion

This is a particular problem with honeycomb cores that causes weight gain, corrosion in aluminum honeycomb, and disbonds if water freezes and expands. It is a very common problem in high-temperature repairs: the heat of curing the repair causes trapped water to turn to steam, disbonding face sheets around the repair and ultimately converting a small area of damage into a large one.

Lightning, Fire and Heat Damage

Lightning damage ranges from simple surface abrasions to large through-holes, and often includes thermal damage to resin. Fire or excessive heat may char or burn resin and/or fibers. It may be difficult to determine the extent and depth of this type of damage, but it is extremely important to do so. The following are indications of heat and lightning damage: discolored paint, blistered paint, missing paint around fastener holes, missing paint, frayed or loose fiber.

Damage Detection

Damage to composites is often hidden to the eye. Where a metal structure will show a dent or "ding" after being damaged, a composite structure may show no visible signs of it and yet may have delaminated plies or other damage within.

Impact energy affects the visibility, as well as the severity, of damage in composite structures. High and medium energy impacts, while severe, are easy to detect. Low energy impacts can easily cause hidden damage (See Figure 14 - 4, differences in damage visbility and severity resulting from medium and low energy impacts). There are a variety of non-destructive inspection techniques available to help determine the extent and degree of damage.

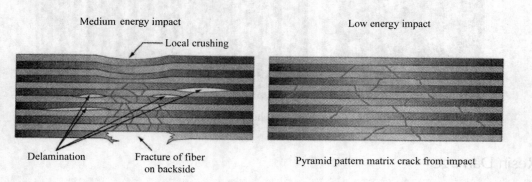

Figure 14 - 4　Difficulties in damage detection

Component Identification

Ideally, composite components should be fully identified before a repair is performed. Such details as material specifications, number of plies and ply orientations, core ribbon direction, ply buildups and drop-offs, and numerous other items need to be clarified and understood before beginning a repair.

For aircraft, this type of information is often available in the structural repair manual (SRM) or equivalent documents. Even if an SRM or equivalent is not available, this type of information is still needed for proper repairs. Determination can often be done by careful taper sanding through a small sample of the damaged part, and reading the information directly from the composite itself. However, a thorough understanding of composite materials — including weave patterns and ply orientation concepts such as balance, symmetry, nesting, etc. (Figures 14 - 5 and 14 - 6).

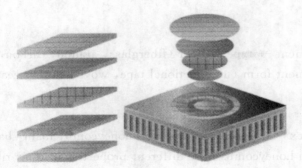

Figure 14 – 5 Component identification

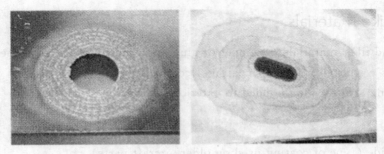

Figure 14 – 6 Component identification by careful taper sanding

Repair Materials

A repair should use the same materials, fiber orientation, core orientation, stacking sequence, nesting, curing temperature and sealing as was used in the original fabrication of the part. However, sometimes this is not feasible, especially in the case that an original structure was manufactured with a fiber/resin system that required an autoclave cure, and the repair is being performed without using an autoclave. Therefore, material substitutions may be allowed or even required by the SRM. In "battle damage" or temporary repairs, often more flexibility in material substitution is allowed than would be normal for permanent repairs.

Nevertheless, the goal remains as much as possible to return the structure to its original strength, stiffness, shape and surface finish, etc. In any composite repair, the following types of repair materials need to be considered and evaluated before the repair is begun, to ensure that the repair is structurally sound:

Matrix Resins

1. Resin systems — wet layup vs. prepreg, low-temperature vs. high-temperature.
2. Cure cycle requirements and available equipment.

Fibers/Fabrics

1. Fiber reinforcement — type of fiber (fiberglass, aramid, carbon, etc.).

2. Fiber reinforcement form (unidirectional tape, woven cloth, weave style, etc.).

Core Materials

1. Core type-Nomex® honeycomb, aluminum honeycomb, foam, balsa, etc.

2. Core orientation-honeycombs have different properties in their ribbon and transverse directions.

3. Core adhesive or potting compound.

Lightning Strike Materials

1. Mesh and fiber materials — copper or aluminum mesh, conductive fibers, etc.

2. Coatings — nickel or ceramic coatings, conductive paints.

3. Grounding strips (often bonded to exterior surfaces on aircraft).

Sealants

1. Polysulfide (type of coating used on older aircraft parts).

2. Polyurethane — newer systems and hybridized coatings that are less brittle, more damage tolerant.

3. Rubberized coatings (variety of products to help prevent damage from erosion, etc.).

Paint Removal

The first step is to remove paint and/or outer coatings. Chemical paint strippers must not be used, unless you are certain they are specifically designed for composite structures. Most paint strippers are based on methylene chloride, and will attack cured epoxy resin (which is a common matrix resin for composites). Paint and coatings may be removed by: (1) Hand sanding. (2) Bio-based starch media blasting (wheat, corn, corn hybrid polymers). (3) Careful grit blasting or plastic media blasting.

It is important to check any repair manuals- such as SRMs for aircraft-or guidelines offered by manufacturers, and also to make sure all health and safety requirements are met.

Chemical Stripping

Most paint strippers are methylene chloride-based. This solvent will attack cured epoxy resin, and is not recommended for composite structures. Alter-natives include hand sanding and various particulate blasting methods.

Hand Sanding

Hand sanding is widely used as a paint removal method for composites. No expensive equipment is needed, and the sanding pressure can easily be controlled to avoid laminate damage. Hand sanding is recommended for paint removal in a small area (for example: in an area to be repaired). One must be careful not to damage fibers in the surface ply, unless a specific sanding ply has been included on the surface of the component. The obvious disadvantage is the high labor costs, especially on large parts. However, this is by far the most common paint removal method for repair preparation.

Blasting

There are many types of blasting in use today: (1) Sandblasting. This type of blasting is possible with composites but must be done very carefully, as it is highly likely that fibers on the top ply will be damaged. (2) Aluminum Oxide Blasting. Similar to sandblasting, and the same precautions apply. (3) Plastic Media Blasting. This method is less aggressive than sand or aluminum oxide blasting, and is used often with composites. While effective, this technique has shown some difficulties in the field. The plastic media is most often cleaned and re-used in such systems. This can lead to two problems: 1) The cleaning process is largely designed to remove paint chips and solid matter. If the plastic media becomes contaminated with oil, grease, fuel, etc. , these contaminates can be driven back into the composite surface, causing paint adhesion problems. 2) The plastic particles become dull with reuse, and a poorly trained operator may turn up the air pressure to compensate. This can lead to damage of the composite surface, or even blowing a hole through these rather brittle materials. (4) Starch Media Blasting. This method uses starch media to blast paint from parts. Starch is much more aggressive when used at higher pressures (over 30 psi). The particles fracture upon impact into smaller particles with increased surface edges. These are even more effective at removing coatings, but can also damage composite substrates. A lower blasting pressure and smaller angle of impingement are recommended for composites, and the media flow range and blasting standoff distance should be carefully selected for the part being treated. With starch media, there is no chemical attack to worry about, and it is an environmentally benign and biodegradable material that can easily be disposed of. Although very rare and limited to specific areas within the process, dust generated from the wheat starch can possibly explode. This is not possible when standard blast procedures and ventilation are used.

Other Methods

There are many other methods, such as dry ice blasting, Xenon flash lamps, baking soda blasting, etc. Work in this field is continuing, and better methods are currently being researched by many organizations. Regardless of the blasting system used, one development that is universally accepted is the preference for a flat nozzle with a fan-shaped blast path. This avoids the damaging hot spot in the blast path center of traditional round nozzles.

Damage Removal

After paint removal, additional damage assessment is performed, because the hidden damage now becomes more apparent. All damaged material must be removed, including anything contaminated by water or other fluids and anything showing signs of corrosion. Be sure to inspect aluminum honeycomb for corrosion. Beginning signs of corrosion include a dull instead of a shiny appearance, and white or gray-colored areas. In advanced corrosion, the honeycomb becomes brittle and tends to flake. Remove the damaged plies, make a smooth circular or oval shape. Take care not to damage the undamaged plies.

Repair Design

Designing a composite repair is often quite a complex process. This can be seen in Table 14 – 1, which lists some of the considerations, practicalities and parameters of composite repair.

Table 14 – 1　Composite Repair Design Considerations

Design considerations	Practicalities	Parameters
• Temporary or permanent • Bonded or bolted • Flush or external doublers • Single-sided or double-sided • Wet lay-up or prepreg • Appearance desired • Access to part and/ or damaged area • Details (shape, heat sinks, etc.)	• Time • Environment • Equipment • Skills • Materials • Regulatory and organizational requirements • Failure consequences (cosmetic fairing or primary structure)	• Strength • Stiffness • Weight • Shape/Contours • Appearance/ Surface finish

The perfect repair is to replace the damaged part with a new one. If replacement is not possible, then the ideal repair is to match all original design parameters exactly (e. g. materials, fiber orientation, curing temperature, etc.). In reality, this is rarely possible and compromises are inevitable. However, the design goal of the repair is to return the structure, as much as possible, to its original strength, stiffness, shape and surface finish, etc. If the extent of damage or overall damaged area exceeds allowable repair limits in applicable repair manuals, then specific engineering support is required in order to proceed with the repair. An individually designed repair is necessary. Since it is such a complex subject, specific training in composite repair design is a must.

Vacuum Bagging Materials for Composite Repair

How to sequence and apply the variety of materials used in vacuum bagging can be one of the most confusing aspects in performing composite repair. The sequence of materials used is called the vacuum bagging schedule (or simply, bagging schedule).

Vacuum Bagging Requirements for Repairs

To obtain a good vacuum, certain requirements must be met:

1. The vacuum bag and the base plate, tooling or mold must not be porous — it must be airtight, hold a vacuum, etc.

2. Leaks must be eliminated or reduced within a specified range.

3. The vacuum pump must be large enough to accommodate the volume of the bagged area.

4. A vacuum gauge should be used on the repair to check that the required vacuum is achieved.

The breather is another important factor in achieving good vacuum. The basic rules are:

1. There should be a continuous breather layer across the area being vacuum bagged (Figure 14 – 7. A typical bagging schedule shows the routes for extraction of excess resin (orange arrows) and gases (blue arrows.))

2. The breather should not fill up with resin.

3. The breather and bleeder layers should make contact at the edges.

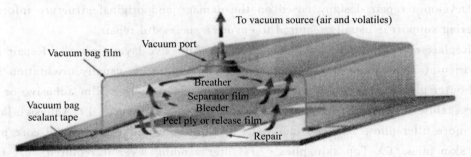

Figure 14 – 7   Bleeder, breather, and bag sequence

If both the bleeder and breather layers fill up with resin, the dynamics beneath the vacuum bag change. Instead of having atmospheric pressure exerting down-force on the laminate consolidating the plies, the result is a mass of liquid resin under pressure. In this case, the plies are free to float within the resin mass, rather than being consolidated downward toward the tool face. For this reason, a separator layer is often used between the bleeder and breather layers, so that the excess resin being pulled into the bleeder layer does not move up and saturate the breather layer.

Repair Instructions

Below is a review of the general steps in performing a composite scarfed repair with a flush, bonded repair patch.

1. Get best access possible, both sides if feasible.

2. Inspect for extent of damage: (1) Visual; (2) Tap, ultrasonic, X-rays, etc.

3. Remove all damaged and contaminated material. (1) Remove damage in a circular, oval or rounded pattern. (2) Remove or treat contaminated material.

4. Determine the part's ply count, orientations, laminate thickness and materials in preparation for repair design and scarfing. There are several choices depending on what information and materials are available. (1) Thorough repair instructions will detail exactly the number of plies, ply orientations and material specifications for the repair, as well as the scarf diameter for the repair area. (2) Ply count, ply orientations and material specifications may also be found in documentation of the original structure's manufacture (engineering drawings). (3) To determine plies, orientations and material practically, there are 2 options: 1) If a large enough piece of the damaged material is available, scarfing into that will reveal the number of plies, etc. 2) Otherwise, it will be necessary to taper sand into the area around the damage, roughly 1/2 inch for each ply of laminate thickness. Use calipers to determine the laminate thickness.

5. Taper sand/scarf the repair area, according to repair design instructions, to create a smooth, flat surface with high surface energy.

6. Thoroughly dry the structure if moisture is present.

7. Develop a repair design, based on the damage and original structure information. Engineering support is usually required to ensure a successful repair.

8. Replace materials. (1) Solid laminate: 1) Adhesive layer first. 2) Repair plies — match orientations with original structure. 3) Extra plies — usually orientation matches original outer ply. 4) Outer sanding layer if required (usually a film adhesive or one-ply fiberglass fabric layer). (2) Through-damaged sandwich structure: 1) Adhesive layer. 2) One or more filler plies. 3) Core material. 4) Core splice adhesive around core plug. 5) Bottom skin plies. 6) Top skin plies. 7) Outer sanding layer if required. (A two-step process is recommended for sandwich panel structures where the core is bonded then machined flush, followed by the skin repairs. The top and bottom skin repairs follow solid laminate guidelines above.)

9. Vacuum-bag and cure repair plies as required.

10. Inspect repair.

11. Sand and finish as required. Do not sand into fibers of repair plies.

Scarfing

After completing initial damage removal, the area around the repair must be prepared. The corners of the damage removal hole must be rounded off and the area beyond this should be tapered to provide the best load transfer when the repair patch is bonded in. Scarfing, or taper sanding, is usually achieved using a compressed-air-powered high-speed grinder. This is a gentle process, which prepares the damaged area for application of a repair patch. It is imperative to follow all repair manual guidelines, and significant skill and practice on the part of the repair technician is mandatory.

If the damaged area exceeds the allowable repair limits in the governing repair manuals, then specific engineering support is required in order to proceed with the repair. Tapered (scarf) repairs are typically circular or oval. There are two ways to specify how much area should be tapered around the damage: (1) tapered-scarf distance per ply, or (2) tapered-scarf angle.

The Repair Patch

The plies of the repair patch are cut to fit the prepared repair area and should have rounded corners. The repair patch is attempting to replace the damaged area in the composite laminate exactly, restoring it as much as possible to original. Thus, the number of plies and orientations of each ply must match, layer for layer, that of the original structure. Each ply in the repair patch matches the corresponding layer in the scarfed repair area.

Therefore, in an exact ply-by-ply replacement, will the repaired structure be as strong as the original? No. Depending on many details, the repaired structure is typically about 70%–80% as strong as the original undamaged structure.

Is it possible to make a repaired structure as strong as the original? Yes. However, extra repair plies must be added to compensate for the loss of strength caused by the repair. This means the repair will not be perfectly flush, and also that the repaired structure will be stiffer than the original.

Is the extra stiffness a problem? If the original structure is not stiffness-critical and is primarily loaded in straight tension or compression, then a stiffer repair will most likely be fine. However, if the structure flexes significantly under load, the stiffened area of the full-strength repair may contribute to failure of the repair. Some repairs may therefore need to be deliberately under-strength, in order to match the stiffness of that original structure.

Curing Methods

Hot Bonders

Portable repair equipment is commonly referred to as hot bonders. Produced by a variety of manufacturers, most hot bonders are suitcase-sized or smaller and are used to control the application of heat and vacuum to a composite repair. They are especially useful for field repairs, in situations where it is not possible to remove the damaged part for repair. They also can be used to monitor and control temperature in oven-cured repairs.

Hot bonders are most commonly used to control heat blankets, but can also be used to control heat lamps, hot air guns, or even ovens. They can be programmed to store as many as 30 different cure cycles in memory. They require an electrical power source, and control the heating through thermocouples.

The cure parameters form the cure cycle, which is pre-programmed into the hot bonder. Most hot bonders can store multiple cure cycles in memory. Operators select the cure cycle desired by entering a number on the keypad, similar to speed-dial. Cure cycles tell the hot bonder how fast to warm up (ramp), how long to hold the cure temperature (soak), and how fast to cool down. Cure cycles will also specify vacuum pressure. Cure cycles for hot bonder repairs typically have a single soak as opposed to those for autoclave repairs which often have two soaks. Apart from controlling the temperature of a repair the hot bonder is also necessary to control the amount of vacuum applied.

Heat Blankets and Other Heating Methods

Accurate temperature control and uniformity are difficult with heat lamps and hot air guns. Temporary ovens and heat blankets offer better control of curing temperature and even distribution of heating. Heat for curing the repair is supplied from a silicon rubber heater blanket. It consists of a heating element sandwiched between two layers of silicon rubber sheet. Being silicon rubber the sheet will not stick if it comes into contact with the resin and will withstand high temperatures.

A heat blanket the same size as the repair area is too small. Heat blankets must be considerably larger than the area being cured. For example, an 8-inch diameter circular repair will require at least a 12-inch diameter heat blanket. Using this size of blanket allows for thermocouple placement well inside the blanket edges.

Controlling thermocouples placed near the edge will make the blanket overheat, causing a very real fire risk and a damage risk to the component. Also, temperatures drop within 2 inches of the blanket's edge. At the very edge of the blanket, the temperature is easily 100°F colder than at its center. This is often referred to as the heat blanket's "cold zone."

Cure Temperature Considerations

There are four common choices, with variations, as shown in Table 14－2.

Table 14－2　Common Cure Temperatures

Curing temperature	Manufacturing method
Room temp －77°F (25℃)	Wet lay-up
Room temp with post-cure at 150°F (66℃)～200°F (93℃)	Wet lay-up
250°F (121℃)	Prepregs
350°F (187℃)	Prepregs

Which is best? It depends. Post-cures are very often required for room-temperature laminating resins, in order to develop full strength in a reasonable time. Often, but not always, prepregs cured at higher temperatures are stronger than room temperature cured materials. However, high curing temperatures can cause problems:

1. Steam creation in laminates and over cores.
2. Blown skins.

3. Excessive porosity in bondlines.

4. Uneven heating problems.

5. Overheating/fire risks.

6. Increased documentation, often with expensive equipment.

7. More training required.

Often, repair manuals will offer a choice, especially with smaller repairs. In this situation, go with the lowest-temperature repair allowed by the manual or guidelines. It will offer an easier repair with less ways for things to go wrong. If there is no choice, then it is imperative to do exactly what is prescribed in the manual or guidelines.

Thermocouple Placement and Other Issues

Proper thermocouple placement is crucial for proper high-temperature repairs using hot bonders. When heat blankets are used, multiple thermocouples are required. The general guideline is: the more thermocouples, the better. This is because thermocouple failures are common and replacement during a running cure is virtually impossible. Another reason for using numerous thermocouples is to control temperature spread across the repair. Remember, the goal is to supply uniform heat to the repair and avoid high spikes and/or cold spots. Temperatures can vary widely across a repair during cure.

New Words/Phrases/Expression

1. intrusive[ɪn'trusɪv, -zɪv]　*adj.* 闯入的；打扰的；侵入的

2. implement['ɪmpləmənt]　*vt. & n.* 实施；执行；使生效；实现；落实；把……填满；工具；器械；家具；手段；履行

3. vacuum['vækjuəm]　*n. & v.* 真空；空白；空虚；清洁；用真空吸尘器清扫

4. scarf[skɑːrf]　*n. & v.* 嵌接；围巾，领巾；桌巾，台巾；围（围巾）；打（领带）；披（披巾）；用围巾围

5. susceptible[sə'sɛptəbəl]　*adj.* 易受影响的；易受感染的；善感的

6. damage['dæmɪdʒ]　*v. & n.* 损害，毁坏；损伤，损害，损毁；赔偿金

7. puncture['pʌŋktʃə]　*vt. & vi. & n.* 刺，戳；贬低，削弱；揭穿；(指轮胎等)被刺穿；刺痕；(车胎等的)刺孔

8. impact['ɪmˌpækt]　*n.* 碰撞，冲击，撞击；冲击力

9. moisture['mɔɪstʃə]　*n.* 水分；湿气；潮湿；降雨量

10. fluid['fluɪd]　*n. & adj.* 液体，流体；流体的，流动的，液体的；易变的，不固定的

11. ingress['ɪnˌgrɛs]　*n.* 进入；进入权；进食

12. facilitate[fə'sɪlɪˌtet]　*vt.* 促进，助长；使容易；帮助

13. freeze-thaw cycling　冻融循环

14. stripper['strɪpə]　*n.* 褪去剂，剥皮剂，脱衣舞表演者，脱衣舞女

15. intrusion[ɪn'truʒən]　*n.* 闯入；打扰；干扰；干涉

16. char[tʃar]　*v.* & *n.* 把……烧成炭，把……烧焦；烧焦；碳；（尤指家庭的）杂务

17. lightning damage　雷击损伤

18. orientation[ˌɔriənˈteʃən]　*n.* 方向，定位，取向，排列方向

19. ply[plaɪ]　*n.* 股；层；厚；（夹板的）层片

20. autoclave[ˈɔːtoʊkleɪv]　*n.* 压热器；高压蒸气灭菌器

21. grounding strip　接地片，接地带

22. sealant[ˈsilənt]　*n.* 密封剂

23. polysulfide[ˈpɒliːsʌlfaɪd]　*n.* 多硫化物

24. sandblast[ˈsændblæst]　*n.* & *v.* 喷沙，喷沙器；喷沙

25. hot bonder　热粘合机，热焊机

26. cure[kjʊr]　*n.* & *v.* 治愈；药物；疗法；措施治愈；矫正；解决；消除；被加工处理；固化

27. wet lay-up　湿铺层

28. prepreg[ˈpriːpreg]　*n.* （塑料或其他复合材料在模塑之前用树脂浸泡的）预浸料坯

29. thermocouple[ˈθɜːməkʌpəl]　*n.* 热电偶

30. curing agent　固化剂

Notes

（1）The damage evaluation process（DEP）is a step-by-step procedure to make an analysis of the damage to structure and find the correct action.

翻译：损伤评估流程（DEP）是一步步地对结构损伤进行分析从而找出正确修理措施的过程。

（2）Chapter 5 of the maintenance manual shows inspections for damage incidents that can cause hidden damage that is more than the structure was first designed for.

翻译：AMM 05 章节列出了发现意外损伤后的检查工作，意外损伤可能造成远比最初设计出的结构损伤多的隐藏损伤。

（3）Apart from controlling the temperature of a repair the hot bonder is also necessary to control the amount of vacuum applied.

翻译：除了控制修理的温度，热黏合机同时也提供了必需的压力控制。

（4）Heat for curing the repair is supplied from a silicon rubber heater blanket. It consists of a heating element sandwiched between two layers of silicon rubber sheet. Being silicon rubber the sheet will not stick if it comes into contact with the resin and will withstand high temperatures.

翻译：硅胶热电毯提供了固化修理所需的热量。它由两层硅胶片夹着一层发热元件构成。硅胶片不会太硬所以它能更好地与结构表面的树脂相接触且能抵挡高温。

（5）The following are indications of heat and lightning damage：discolored paint，blistered paint，missing paint around fastener holes，missing paint，frayed or loose fiber.

翻译：热损伤和雷击损伤迹象如下：漆层褪色、漆层起泡、紧固件孔周围掉漆、漆层丢失、纤维磨损或起毛。

(6)Remove the damaged plies，make a smooth circular or oval shape. Take care not to damage the undamaged plies.

翻译：去除损伤的纤维铺层，去除部位为圆角平滑的圆形或椭圆形。注意不要损伤未受损的纤维铺层。

Exercises

Ⅰ. Direction：choose the right meaning for each item.

（1）织物预浸料

A. fabric impregnation　　　　　B. fabric preform　　　　　C. fabric prereg

（2）凝胶涂层

A. gel coat　　　　　　　　　　B. gel drum　　　　　　　　C. gel flow

（3）上漆

A. lacquer formation　　　　　　B. lacquer finish　　　　　　C. lacquer putty

（4）加热固化

A. oven heating　　　　　　　　B. over heating　　　　　　C. over flow

（5）涂漆缺陷

A. paint blower　　　　　　　　B. paint dilution　　　　　　C. paint blemish

（6）树脂配方

A. resin formulation　　　　　　B. retaining nest　　　　　　C. reactive resin

（7）湿砂打磨

A. sandblasting　　　　　　　　B. sand wet　　　　　　　　C. starch blasting

（8）真空袋成型

A. vacuum assisted resin injection

B. vacuum bag molding

C. resin transfer molding

（9）玻璃纤维缠绕机

A. glass fiber winding machine　　B. glass wool　　　　　　　C. glass yarn

（10）粘连织物

A. bonded fabric　　　　　　　B. knitted fabric　　　　　　C. woven fabric

（11）损伤容限

A. damage threshold　　　　　　B. damage tolerance　　　　C. damage limit

（12）透气毡

A. breather blanket　　　　　　B. breather film　　　　　　C. breather cloth

（13）铺层对接

A. butt lay-up　　　　　　　　B. butt line　　　　　　　　C. butt wrap

（14）胶疙瘩

A. catenary　　　　　　　　　B. cat eyes　　　　　　　　C. catalyst

（15）净化间

A. clean room   B. clean up   C. cleaner

（16）耦合剂

A. coupling medium   B. coupling agent   C. coupling film

Ⅱ. Fill the right information in the form.

No.	Term	Abbreviation	Chinese
（1）	Automated Fiber Placement	AFP	
（2）	Bond Assembly Fixture	BAF	
（3）	Bulk Molding Compound	BMC	
（4）	Ceramic Matrix Composite	CMC	

Ⅲ. Fill in the following blanks according to the text.

（1）A repair should use the same materials, _____, core orientation, _____, nesting, _____ and sealing as was used in the original _____ of the part. However, sometimes this is not _____, especially in the case that an original structure was manufactured with a fiber/resin system that required an _____ cure, and the repair is being performed without using an autoclave. Therefore, material _____ may be allowed or even required by the SRM.

（2）After paint removal, additional damage _____ is performed, because the hidden damage now becomes more _____. All damaged material must be _____, including anything contaminated by water or other fluids and anything showing signs of _____. Be sure to inspect aluminum _____ for corrosion. Beginning signs of corrosion include a ____ instead of a _____ appearance, and white or gray-colored areas. In advanced corrosion, the honeycomb becomes _____ and _____ to flake.

Ⅳ. Answer the following questions.

（1）What is a delamination?

（2）What are the repair instructions of the composite structure?

（3）Please describe the component identification of the composite repair.

（4）How many methods are used in the paint and coatings removal of the composite materials?

（5）What is the goal of the composite structures repair?

（6）In an exact ply-by-ply replacement, will the repaired structure be as strong as the original?

（7）Is it possible to make a repaired structure as strong as the original?

（8）Is the extra stiffness a problem?

（9）Which cure temperature is the best?

（10）What are the considerations should be considered in composite repair design?

Ⅴ. Translating the following sentences into Chinese.

（1）Perhaps the most common of all damage types, delaminations are often caused by low-energy impacts, such as a tool drop or a glancing bird strike on an aircraft. They can sometimes be visible if they are near the surface. Delamination can also be caused by manufacturing defects, resin starvation in a laminate and by moisture ingression.

(2) Lightning damage ranges from simple surface abrasions to large through-holes, and often includes thermal damage to resin. Fire or excessive heat may char or burn resin and/or fibers.

(3) Ideally, composite components should be fully identified before a repair is performed. Such details as material specifications, number of plies and ply orientations, core ribbon direction, ply buildups and drop-offs, and numerous other items need to be clarified and understood before beginning a repair.

(4) The perfect repair is to replace the damaged part with a new one. If replacement is not possible, then the ideal repair is to match all original design parameters exactly (e. g. materials, fiber orientation, curing temperature, etc.). In reality, this is rarely possible and compromises are inevitable.

(5) If the damaged area exceeds the allowable repair limits in the governing repair manuals, then specific engineering support is required in order to proceed with the repair.

EXTENSIVE READING

Damage to composites

Because of their complexity, the fracture mechanics of composites are drastically different compared to metals. The fracture behavior of the composite depends on the properties of the two components as well as the interaction between them. The modes of fracture and failure also depend on the lay-up and geometry of the composite, impact load and velocity, loading direction, and environmental factors. In one study, it was found that increases in aging time, aging temperature, and oxygen concentration degraded the mechanical characteristics of the composite. Another study showed that the number of dissimilar interfaces in the lay-up of carbon fiber epoxy composites changes the energy needed for delamination to occur. With all of these variables, the types of damage and failure in composites are more diverse and involved, with some occurring concurrently. When a composite is impacted by a projectile, shear and normal stresses develop within the composite. Depending on the type and magnitude of impact, the resulting stresses will cause a progression of damages within the composite. This typically consists of matrix cracking, followed by delamination and finally fiber breakage at higher loads. Matrix or interlaminar cracking occurs first at low level impact energies. As the severity of the damage increases, delaminations or cracks between the plies will develop, significantly lowering the strength of the composite. At high levels of damage, the fiber fracture will occur, causing severe and potentially irreversible damage to the composites. Other types of composite damage include but are not limited to fiber/matrix debonding, fiber buckling, and matrix crushing.

As stated, the extent of resulting damage depends heavily on the energy of the impact, shown in Figure 14 - 8. High energy impacts, usually caused by high speed projectiles, will cause surface and penetrating damage that is visibly detected, often requiring extensive repairing or complete replacement of the component. By comparison, low energy impact, considered in the following chapters, may only cause internal defects, such as delaminations, that can be detected solely through nondestructive evaluation techniques, hence giving it the name Barely Visible Impact Damage or BVID. For a composite system, the critical impact energy marking the transition from internal to visible damage can be found through impact test studies. In a study by Sultan et al, three critical impact energy loads were determined, marking where the dominating fracture mode changed from matrix cracking to fiber cracking to major damage. In another study, the impact energy applied to graphite epoxy composites was found to cause fiber breakage once it reached 15 Joules.

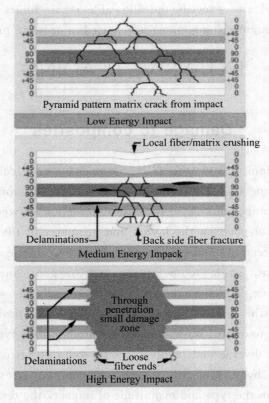

Figure 14 - 8　Types of damage in composites

Delamination, or the formation of internal cracks between plies, is one of the most common fracture modes in composites and when undetected, can result in catastrophic failure of the composite. Besides low energy impacts, delamination is typically caused by the development of stresses between plies that can be caused by manufacturing defects and environmental effects. Defects typically occur during the manufacture of the composite part or when the composite component is being assembled. There are many types of defects including voids, incomplete resin cure, tolerance errors, debonding, and foreign body inclusions. As discussed previously, environmental effects, such as temperature gradients during curing and moisture absorption can cause residual stresses that may cause delamination and debonding.

Delaminations can be detrimental to the strength of the composite, with a reduction in compressive strength of up to 50%. The compressive strength has been shown to be significantly more affected than the tensile strength. Delaminations caused by low energy impact resulted in no change in the tensile strength, but compressive strength evaluated through compression after impact was only 40% of its original strength. Delaminations also cause other structural effects such as stress concentrations, stiffness loss, and possible failure of the part. In one study, bending tests were performed on delaminated carbon fiber epoxy panels. After several millimeters of displacement, the delaminations began to propagate and a drastic drop in force was observed, simulating the failure of the composite. Therefore thorough detection is needed to locate delaminations and defects that could lead to composite failure. Detection methods include ultrasonic scanning, low frequency vibration methods, and thermography, several of which are used. Delamination growth is governed by the three major fracture modes and mixed mode combinations. By analyzing the energy release rates and stress intensity factors of all three modes, one can determine the stress state at the crack tip and therefore predict delamination. Determining the critical interlaminar fracture toughness and using it in conjunction with fracture analysis can also help researchers understand delamination behavior. Fractographic analysis can be used to identify fracture features and increase understanding of fracture modes, as shown in several studies of carbon fiber composites. Various methods can be used to help prevent delaminations and their propagation, such as selection of the matrix and reinforcement components. For example, it was found that switching the matrix from epoxy to bismaleimide resulted in a toughening effect that increased the strength of the matrix and created a stronger interface bond between the matrix and fibers. Incorporation of a filler into the matrix, as stated previously, can also make the composite tougher and resistant to delamination. Finally, performing a surface treatment on carbon fibers has been shown to strengthen the fiber-matrix interface, leading to an increase in interlaminar fracture toughness and resistance to delamination. But with the constant risk of delamination, methods must be investigated to temporarily or permanently repair damaged components.

Repair methods

Depending on the severity and type of damage, there are several ways to approach the repair of a damaged composite part. For widespread and severe damage, the composite part may have to be removed and completely replaced, an expensive and timely procedure. However, for smaller, more localized damage, it is advantageous to isolate and repair the damaged area to save time and money as well as maintaining the integrity of the structure.

Patch repair

One common repair method is the use of composite patches. A patch, typically made from composites, is bonded over the damaged area. The patch is used to transfer loads throughout the part and in the process avoid applying further stress to the damaged area. Patch repairs typically require limited surface preparation, are easy to disassemble, and are typically mechanically bonded with screws. However fasteners often cause stress concentrations that can cause further damage, making adhesive bonding advantageous. Though effective at transferring loads, careful surface preparation must be made for adhesive bonding, and pressure may be needed to create the bond, potentially damaging the part further. The selection of surface treatment must also be considered, as some, such as grit blasting, may cause surface damage and adversely affect the bond strength. Though simple, the composite patch simply covers the damaged area and internal damage such as delaminations remains unrepaired. The patch also sits above the line of the structure's surface, which may affect the aerodynamics of the composite.

Scarf repair

A more aerodynamic method, similar to the patch repair, is the scarf repair. In a scarf repair, the damaged area is removed or "scarfed" and an angular patch is manufactured to directly fit the area. Though considerably more involved, scarf repairs provide a less expensive alternative to the replacement of thick primary components by producing more reliable and aerodynamic repairs. Scarf repairs have also been shown to give additional strength to the composite. In a study by Harman and Rider, the compression after impact strength of the pristine laminate was only 70% the strength of the scarf-doubler repair joints, showing that the repair provided additional strength to the composite. Compared to patch repairs, a proper scarf repair will eliminate adhesive strength concentrations and have a higher strength. Although effective, this method has several drawbacks. As with the typical patch, the scarf repair does not directly repair internal damage. The angled scarf patches are also difficult to manufacture, require the removal of undamaged composite material, and are often prone to creep due to shear stresses. Scarf repairs are often not performed on thin components because the angles cannot be machined properly. Therefore it is desirable to find a more stable and direct repair method.

Injection Repair

The method investigated in the present work is the injection repair method, shown in Figure 14 - 9. In contrast to the overlaying of the damaged area in the patch and scarf repairs, the injection repair process injects a low viscosity resin into the damaged area. Once the low viscosity resin is injected, the resin spreads throughout the delaminated area, infiltrating the microcracks and repairing the panel internally. The injected panel is then subjected to a high temperature cure schedule to create a hardened crosslinked resin. This method is commonly used for minor damage consisting primarily of delaminations and debonding within the composite. The injection repair method can also be used in conjunction with patch methods to repair severely damaged components. Unlike the previously discussed methods, the injection repair directly repairs the damaged area rather than simply covering the damage. Injection repairs also help to retain the aerodynamics of the composite part. Like any repair method, the injection repair technique must be evaluated and validated on each type of composite to prove its effectiveness.

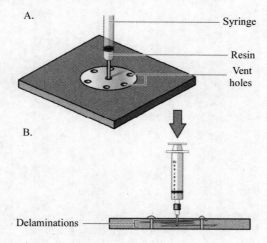

Figure 14 - 9   Injection repair

As with the requirements for the composite, the injection repair must be able to withstand the environmental conditions of the application. An unstable repair could result in further damage to the composite part. The repair should also have sufficient strength and stiffness, potentially recovering the strength lost from the pristine composite. The injection repair process has proven successful in restoring strength to the composite in other studies. The injection repair technique was found to achieve a repair efficiency of 98% on graphite epoxy composites when all delaminations were filled. Injection repair has been successfully used on applications other than composites, including increasing the stiffness in cracked reinforced concrete and repairing minor cracks in car windshields.

The selection of the injection resin is critical to the quality and effectiveness of the repair. The resin should meet the following requirements: First, a low viscosity achievable under a wide range of conditions is desired to achieve quick and complete infiltration into the delaminations. Second, to be stable at the high temperatures withstood by aerospace structures, the resin must have a glass transition temperature (T_g) exceeding the maximum service temperature. The combination of a low viscosity and high T_g has been a perpetual problem in finding suitable resins for the injection repair. Third, the resin should achieve a strong adhesive bond with the substrate upon curing to ensure a quality repair. Finally, since the repair may be semi-permanent, the resin should be stable and have low toxicity. Selection of a resin without any of these qualities may fail to repair the composite and cause further damage. Epoxy resins are commonly used for injection repairs, but their low T_g prevents the use of injection repairs for high temperature environments. Demonstrating the use of a resin with satisfactory viscosity and T_g requirements would expand the application of injection repairs to more vigorous environments.

Figure 14.5 Injection repair

Lesson 15　Cabin Layout and Emergency Equipments

INTENSIVE READING

Cabin Layout and Cabin Comfort

An airline's profit depends upon each of its aircraft carrying as many passengers as possible. The less cabin space devoted to passenger seating, the more expensive the air ticket will be. The designers of aircraft are therefore required to accommodate the largest number of seats, while retaining reasonable comfort.

When long-distance air travel became established, after 1930, aircraft were fitted with sleeping compartments. As late as the 1950s, the Douglas DC-7, which operated the first regular non-stop transatlantic services in 1956, taking about ten hours, was fitted with "slumber" reclining seats and bunks. These were not needed in later, faster jets.

The economic significance of high-density seating was soon exploited, and a line of seats on each side of a central aisle became the classic configuration; the second-generation jet airliners had rows of five (three plus two) or six (three plus three) seats (Figure 15 – 1).

Not all airliners have alternative economy (coach) and first-class accommodation. Some commuter and shuttle aircraft, charter operations and small island-hoppers have a single class. The distinction is most evident in the different seat layouts (Figure 15 – 2).

In the A380 (Figure 15 – 3), The passenger compartments accommodate the passengers, cabin crew, and their facilities on the main and upper deck. Two staircases between the decks are installed; one straight staircase installed at M1 doors and one curved staircase is installed at the rear of the cabin. The cabins are divided into utility areas and seating areas. The utility areas are located adjacent to the cabin entrances. The passenger seating areas are located between the utility areas. There are three classes of passenger seats in the two deck cabins: first, business and tourist class seats can be installed either on Main deck or Upper deck. The cabin attendant seats are installed for the use by the cabin attendants during takeoff and landing. The seats are foldable and are installed in the areas of passenger/crew doors and the emergency exits. The number of attendant seats depends on the aircraft configuration.

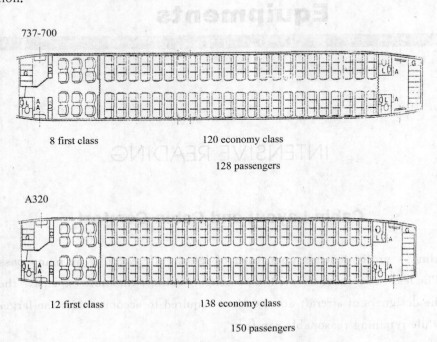

737-700

8 first class

120 economy class

128 passengers

A320

12 first class

138 economy class

150 passengers

Figure 15 – 1　Cabin layout

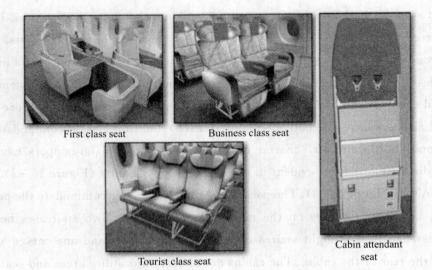

First class seat

Business class seat

Tourist class seat

Cabin attendant seat

Figure 15 – 2　Seat layout

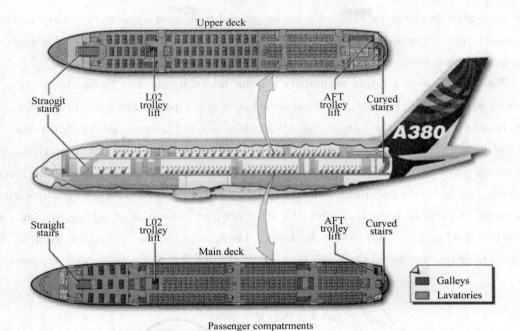

Passenger compatrments

Figure 15 – 3   A380 cabin layout

In narrow-bodied airliners, the economy-section rows of triple-seat units usually give way to two plus two rows of double-seat units in the first-class cabin. In the wide-bodied airliners the standard seating pattern consists of rows of up to nine or ten abreast with two aisles: two plus five plus two in the Douglas DC-10, or three plus four plus three in the Boeing 747. The first-class seating pattern is usually two plus two, with extra luxury provided in the additional lounge areas.

Although an airliner is a parallel tube for most of its length, the cabin tapers at each end, most noticeably in wide-bodied jets, so in the front and back of most aircraft the number of seat rows is reduced. Ultimately the number of seats is determined by the number of seat rows. Seats are usually spaced about 34 inches apart in first class, 32 inches apart in the economy section. This distance, measured from the front of a seat to the same point on the one behind, is known as the seat pitch. To keep fares low, in charter flights, seats may have a pitch of only 29 inches, much closer than for the (more expensive) scheduled services on which an airline's reputation is built.

Most major airlines will permit passengers to select the seats they prefer on mixed-class scheduled flights, as long as they book early enough. Window seats on the side away from the sun give the best view, and seats with most leg-room are by the doors or emergency exits where seat-pitching is most generous.

The handful of seat manufacturers who supply all the world's airlines ensure that the core of comfort provided by adjustable backrests and cushioned upholstery is surrounded by all possible conveniences, plug-in or fold-away meal trays, magazine racks, footrests and luggage restraint devices.

The world's civil aviation authorities set the safety standards. To be able to take the strain of severe air turbulence, or of an emergency stop, passenger seats must be able to withstand several times the normal takeoff acceleration and landing deceleration rates. The UK and US authorities demand that an occupied seat be able to withstand an acceleration of nine times that of gravity (a mass of one pound accelerated by 32 feet per second per second) forward, and one and a half to four and a half times gravity rearward, sideways, upward and downward. Extensive testing ensures the strength and resistance of all welds, bolts and safety-belt anchorages. Adjustable backrests break forward on a sudden stop so that passengers in the seat behind will not hit them if thrown forward on a sudden deceleration.

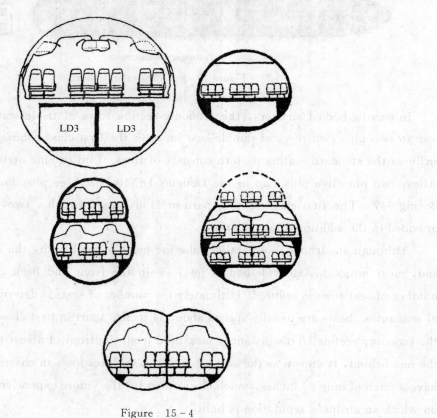

Figure 15 - 4

Most passenger seats face forward, toward the aircraft nose. Backward-facing seats would be safer: passengers would be supported by the backrest during sudden deceleration instead of forward against the seat in front, unrestrained except for the seatbelt fastened across the abdomen.

Because passengers would rather see where they are going than where they have been, airlines have refused to alter this practice but, significantly, whenever military use is made of civil aircraft, seats are turned around into the safer position.

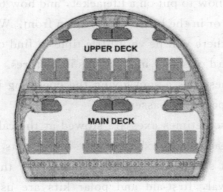

UPPER DECK

MAIN DECK

Figure   15－5

Lifejackets are usually stored beneath each seat where they are readily accessible. Above the passengers' heads, or sometimes released from the seat in front, oxygen masks drop down automatically if the cabin pressure should fall below the optimum level. Seats are upholstered in flame-resistant materials. On some airliners the cushions can be removed and used as floating lift-preservers.

In structure and standard of upholstery there is little difference between first and tourist (coach) class seating. But there is an obvious difference in size. First-class seats are larger and are farther apart, and contoured backrests, often fitted with "slumber" headrests, provide a little extra luxury.

Noise levels vary little between different points on an aircraft, although passengers near a galley or toilet may be disturbed frequently. The engine noise is more noticeable at the back of a rear-engine airliner; the first-class cabin is located away from the engines. From a seat positioned near the landing gears, the sound of the doors locking into the up and down positions can be disturbing, but noise from the wheels on the runway tends to be transmitted throughout the fuselage. Aircraft could be made quieter, but a level of mechanical noise diffuses the sound of 400 passengers talking at once.

Emergency Equipments

Accidents do happen and when they do, crew training, aircraft equipment and the good sense of the passengers may be fundamental to survival. To be alert to (but not obsessed with) the possibility of an emergency increases the chances of surviving it.

Aircraft seat belts, like the seats, are designed to withstand sudden deceleration. Work out how to fasten and unfasten them quickly. Cabin crews check that they are fastened at takeoff and landing, and that babies are installed in special cot-holders.

Do not, unless you are well versed in them, ignore the safety demonstrations at the beginning of a flight. Learn how to put on a lifejacket, and how to use the emergency oxygen masks stored above the seat or in the back of the seat in front. When depressurization occurs suddenly at high altitudes, there will be very little time to find out how to put them on. In each magazine pocket is a card of safety instructions in pictures. It gives the positions of the emergency exits. Each of these has a sign giving the operating instructions in one or more languages, so they can be opened by passengers.

Fire extinguishers and sometimes axes are stowed at the cabin crew stations. Life raft stowages are usually near each main exit; in wide-bodied jets they are extensions of the escape slides, and in aircraft such as the VC10, 707 and 747 they are stored in the ceiling above the aisle. The survival, first-aid and polar kits are usually found near, perhaps attached to, the life rafts or exits.

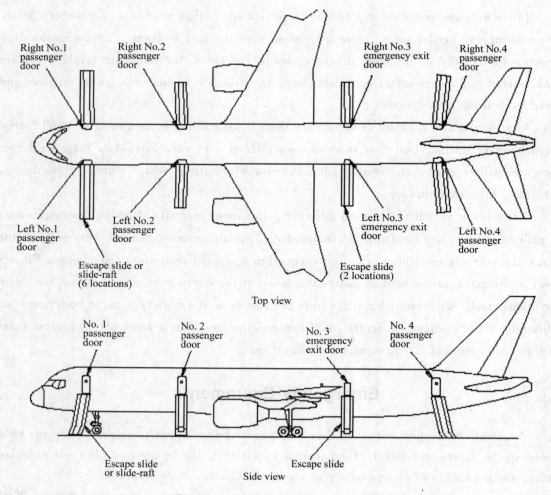

Figure 15-6　Aircrafte excape silides

If the "fasten seat belts" signs flash on in flight, obey them quickly; the aircraft may be about to enter turbulence and passengers walking about, especially at the back, could be thrown around.

In an emergency, obey the cabin crew without question. They will order passengers to fasten seat belts and brace themselves (bracing positions are illustrated on the "safety instructions" card) for sudden deceleration. They man the exits when the aircraft stops, instruct passengers on the procedure, and collect emergency equipment and supplies. If doors are inoperative, they activate the emergency exits, ordering passengers to unfasten seat belts, leave everything behind, and make for the designated escape chutes indicated by the cabin staff.

At the doorway, jump or slide down the chute to the ground (or into the life raft). Then get away from the aircraft. Move fast, and forget your belongings. In a fire, stragglers may be overcome by toxic fumes released by burning cabin furnishings.

Emergency Equipments in A380

Emergency equipment is installed to give to the crew and the passengers all the necessary tools to safely handle all hazardous situations that could occur in the aircraft (Figure 15 - 7).

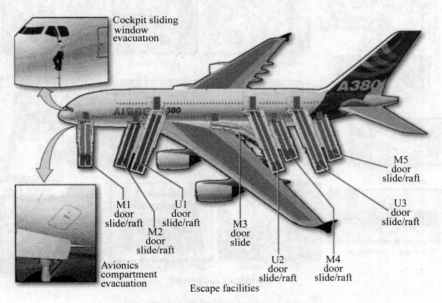

Figure 15 - 7   Escape Facilities

Portable fire extinguishers are installed in the cockpit, in the avionics compartments and in the upper and main deck passenger compartments. Crash axes are installed in the cockpit, in the upper and main deck passenger compartments, to get access to the cabin or out of the cabin in case of an emergency, and to access aircraft components behind panels in case of fire fighting.

Megaphones are installed in the upper and main deck passenger compartments, which can be used if the public announcement system fails or during emergencies.

First aid kits are installed in the upper and main deck passenger compartments. The doctor's kit provides a support to the first aid to passengers and crewmembers when needed.

Emergency locator transmitters are installed in the upper and main deck passenger compartments to make distress alert and location information to Search and Rescue authorities worldwide.

Flashlights are installed at each cabin attendant station and in the cockpit to have portable lights that are independent from the aircraft electrical system.

One manual release tool is installed at each cabin attendant station to give cabin and maintenance crew the means to open or to close the flaps of the passenger oxygen system.

Passenger life vests are installed under each passenger seat or in seat armrests. Cockpit crew life vests are installed in the backrest stowage of the captain and first officer seats and in the life vest stowage of the observer seats. Cabin crew life vests are installed under each cabin attendant seat. Children life vests, spare life vests, demo life vests are installed for the two decks. Life vests give to the passengers means to prevent them from drowning and a way to be recognized (through the color) by search and rescue in the night-light.

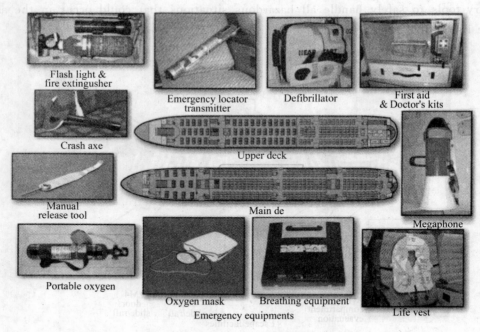

Figure 15 - 8　Emergency equipments

Portable oxygen cylinders are installed in the cockpit and cabin to have additional means for first aid treatment to passengers. A related pressure gauge shall be visible with the cylinder installed. Portable oxygen masks are installed in the cabin to have the means to supply additional oxygen to passengers for first aid treatment. Demonstration oxygen masks are installed in the passenger compartment, evenly distributed between main and upper decks to show passengers how to use portable oxygen masks.

Protective Breathing Equipments（PBEs）are installed in the cockpit and cabin to provide eyes，nose and mouth protection to user against smoke for a duration of 15 minutes.

The defibrillator is used by first responders to cardiac emergencies. An automatic self-testing has been integrated to ensure that the device is ready to go. There is one defibrillator on each deck and it is installed adjacent to medical equipment.

New Words/Phrases/Expression

1. cabin layout　客舱布局
2. high-density seating　高密度座位布置
3. classic configuration　经典布局
4. second-generation　第二代
5. shuttle aircraft　短程（穿梭）机
6. staircase['steəkeɪs]　n. 楼梯;楼梯间
7. first class seat　头等舱座椅
8. business class seat　商务舱座椅
9. tourist（coach）class seat　旅客舱（经济舱）座椅
10. cabin attendant seat　客舱乘务员座椅
11. emergency exit　紧急出口
12. narrow-bodied airliner　窄体客机
13. wide-bodied　宽体的
14. abreast[ə'brest]　adv. & adj 并列,并肩地;并排;并排的;并肩的
15. aisle[aɪl]　n. 过道,通道
16. lounge area　休闲区
17. charter flight　包机
18. scheduled service/flight　定期航班
19. leg-room　（飞机、汽车等）座位前伸腿的空间
20. backrest['bækrest]　n. 靠背
21. cushion['kuʃn]　n. & v. 座垫;垫子;起保护（缓冲）作用的事物；（用垫子）使柔和；（跌倒或碰撞时）起缓冲作用
22. upholstery[ʌp'həulstəri]　n. 家具装饰业;室内装饰品
23. meal tray　小餐架
24. luggage['lʌgɪdʒ]　n.〈英〉行李;〈美〉皮箱
25. safety-belt anchorage　安全带卡头,安全带固定
26. lifejacket['laɪfɪdʒækɪt]　n. 救生衣
27. oxygen mask　氧气面罩
28. flame-resistant material　阻燃材料
29. floating lift-preserver　救生浮筏

30. galley['gæli]　n.(船舰,飞机上的)厨房
31. seat belt　座位安全带;座椅固定带
32. fire extinguisher　灭火器
33. life raft　救生筏
34. escape slide　应急离机滑梯;紧急滑梯
35. survival kit　救生包;救生装备
36. first-aid['fɜːst'eɪd]　adj.急救的,急救用的
37. polar kit　极地服;极地用品包
38. toxic fume　毒烟
39. hazardous['hæzədəs]　adj.&adv.&n.冒险的;有危险的;冒险地,有危险地;冒险,危险
40. portable['pɔːtəbl]　adj.手提的;轻便的
41. avionics compartment　电子舱
42. crash axe　撞击斧
43. megaphone['megəfəʊn]　n.扩音器
44. first aid kit　急救箱
45. emergency locator transmitter　事故位置发射机
46. flashlight['flæʃlaɪt]　n.手电筒;闪光信号灯
47. manual release tool　手动释放工具
48. life vest　救生衣,救生马甲
49. portable oxygen cylinder　便携式氧气瓶
50. Protective Breathing Equipments (PBEs)　呼吸防护设备
51. defibrillator[diːˈfɪbrɪleɪtə(r)]　n.(电击)除颤器
52. cardiac['kɑːdiæk]　adj.&n.心脏(病)的;(胃的)贲门的;心脏病患者;强心剂

Notes

(1)The less cabin space devoted to passenger seating, the more expensive the air ticket will be. The designers of aircraft are therefore required to accommodate the largest number of seats, while retaining reasonable comfort.

分析:devoted to 专心于……;忠于。句中用了"the more ... the more ..."的句型,常表示"越……,就越……"

翻译:机舱用于旅客座椅的空间越少,机票就越贵。飞机设计者们在保证一定的舒适程度情况下,就尽可能多地安置机上座位。

(2)As late as the 1950s, the Douglas DC-7, which operated the first regular non-stop transatlantic services in 1956, taking about ten hours, was fitted with "slumber" reclining seats and bunks.

分析:non-stop 指不经停的(地),不着陆的(地);transatlantic 此处指横跨大西洋的。

翻译:50 年代,1956 年道格拉斯 DC-7 首次开通了跨越大西洋的不着陆的定期航班,飞行经过了 10 个小时,在它上面装有向后倾的座位和卧铺。

(3) Some commuter and shuttle aircraft, charter operations and small island-hoppers have a single class. The distinction is most evident in the different seat layouts.

翻译:有些交通班机、短距来往航班、包机和小型海岛间近距飞行的飞机只有一个等级的座舱。不同的座位布置是最明显的差别。

(4) In narrow-bodied airliners, the economy-section rows of triple-seat units usually give way to two plus two rows of double-seat units in the first-class cabin.

分析:triple-seat 是指三联座椅;double-seat 是指两联座椅;give way to 让步,让某人在先。

翻译:在窄体客机上,经济舱的座椅是每三个座椅连成一体的,而在头等舱中是 2+2 布局的双座位单元。

(5) Although an airliner is a parallel tube for most of its length, the cabin tapers at each end, most noticeably in wide-bodied jets, so in the front and back of most aircraft the number of seat rows is reduced.

翻译:尽管客机机身几乎都是一个沿长度方向的等直径的圆筒,在尾部却逐渐变细,宽体客机的这种变化尤为明显,因而在多数飞机的机头和机尾部分每排的座椅数要减少。

(6) The handful of seat manufacturers who supply all the world's airlines ensure that the core of comfort provided by adjustable backrests and cushioned upholstery is surrounded by all possible conveniences, plug-in or fold-away meal trays, magazine racks, footrests and luggage restraint devices.

分析:handful 少数,少量;一把(的量);be surrounded by 被包围的;关系代词 who 引导的后置定语从句"supply all the world's airlines"修饰先行词 manufacturers;ensure that 后面接从句"the core of comfort provided by adjustable backrests and cushioned upholstery is surrounded by all possible conveniences"作宾语表示要确保的内容;而宾语从句中的主语 the core of comfort 也跟了一个后置定语"provided by adjustable backrests and cushioned upholstery";"plug-in or fold-away meal trays, magazine racks, footrests and luggage restraint devices"作为"all possible conveniences"的同位语进一步解释具体的便利设施。

翻译:只有为数不多的厂家供应了全世界航空公司的座椅。可调的靠背和椅垫保证了座椅主体的舒适性。座椅使乘客得到尽可能的便利,包括可插入的或折叠的餐桌、杂志袋、脚垫和行李固定架。

(7) Backward-facing seats would be safer: passengers would be supported by the backrest during sudden deceleration instead of forward against the seat in front, unrestrained except for the seatbelt fastened across the abdomen.

分析:backward-facing seat 向后的座位;instead of(用……)代替……,(是……)而不是……,(用……)而不用……。

翻译:向后的座位更为安全。因为在飞机突然刹车时,向后座椅有椅背支撑着乘客,而向前的座椅除了腰部系紧的安全带外,没有其他任何固定的装置。

(8) Accidents do happen and when they do, crew training, aircraft equipment and the good sense of the passengers may be fundamental to survival. To be alert to (but not obsessed with) the possibility of an emergency increases the chances of surviving it.

分析:do 此处用于加强语气;the good sense of 良好的意识;be alert to 谨防;be obsessed with 痴迷于……,迷了心窍,受困扰。

翻译:事故的发生是突然的,当它们出现时,机组成员的训练、飞机上的设备和旅客良好的意识可能是幸存下来的基本条件。对事故可能性的警觉(不是烦恼)增加了事故发生时幸存下来的机会。

(9)In an emergency, obey the cabin crew without question. They will order passengers to fasten seat belts and brace themselves (bracing positions are illustrated on the "safety instructions" card) for sudden deceleration. They man the exits when the aircraft stops, instruct passengers on the procedure, and collect emergency equipment and supplies.

分析:without question 毫无疑问;man 作动词,表示"操纵;给……配置人员;在……就位"。

翻译:紧急情况下要无条件地服从乘务人员。他们会要求乘客系好安全带,做好应付突然减速的准备(应急姿势在"安全须知"的卡片上有图示)。飞机停稳后,乘务人员站在出口处,指挥乘客顺序离开飞机并将应急设施和物品收集起来。

(10)At the doorway, jump or slide down the chute to the ground (or into the life raft). Then get away from the aircraft. Move fast, and forget your belongings. In a fire, stragglers may be overcome by toxic fumes released by burning cabin furnishings.

翻译:在门口,跳入滑梯滑向地面(或滑向救生筏中),从而离开飞机。快速行动不要考虑你所带的财物。在着火情况下,落在后面的人可能被燃烧的座舱壁板释放出的有毒气体熏到。

Exercises

Ⅰ. Answer the following according to the instructions after each question.

(1) Tickets for airline seats which take up less cabin space are more expensive.

(True/ False)

(2) A passenger air ticket on a small aircraft will cost more than the same ticket on a large aircraft.

(True/ False)

(3) The potential load factor of a passenger aircraft is the most important factor in the design of that aircraft.

(True/ False)

(4) "The economic significance of high-density seating was soon exploited" means:

(a) Since the greater the number of passengers, the lower the cost of fares and the higher the airline profits, every effort has been made to use aircraft with as many seats as possible.

(b) Airline companies, realizing that the more paying passengers they carried, the greater would be their profits, exploited the situation by getting as many seats into the aircraft as possible, with little regard for passenger comfort.

(c) Airlines, and aircraft designers, have come to understand the importance of economic factors, and these have led them to design and operate aircraft with maximum passenger seating, without necessarily sacrificing comfort and safety.

(Tick the best answer)

(5) Single class airliners are generally restricted to use on short-haul flights, or by charter companies.

(True/ False)

(6) "The handful of manufacturers all possible conveniences" means:

(a) The manufacturers have their hands full to supply all the possible conveniences to the world's airlines.

(b) There are relatively few manufacturers who supply all possible conveniences to the world's airlines.

(c) There are not many manufacturers but they all do their best to supply the essentials of comfort and convenience to the world's airlines.

(Tick one)

(7) Compared with economy class seating, first-class seats are: more relaxed, much safer, more expensive, distinctly different.

(Delete one)

(8) (a) The safety standards for civil aviation are set by an international organization.

(b) Safety standards in civil aviation are set by different authorities throughout the world.

(Tick one)

(9) In Britain and the United States airline passenger seat strength is rigidly defined and carefully controlled by extensive testing.

(True/ False)

(10) Seats are designed to resist deceleration forces (such as might be met in a takeoff or landing crash) higher than forces that might be otherwise encountered (eg. in turbulence)

(True/ False)

(11) (a) Adjustable backrests are designed to break so that passengers thrown forward will not be injured.

(b) The seat backrest can be adjusted, and will, under sudden deceleration, swing forward out of the way of passengers sitting behind.

(c) Seat backrests can be adjusted so that they will break forward out of the way of passengers sitting behind.

(Tick one)

(12) In the two paragraphs "Most passenger seats ... into the safer position", the text says, amongst other things, that backward facing seats:

(a) are safer — as their use by the military implies — but have little passenger appeal.

(b) might be safer if more passengers and airlines could be persuaded to install them as the military do.

(c) are more attractive to civil passengers who, unlike soldiers, like to know where they are going, which is more important to them.

(Tick one)

(13) (a) Engine noise is more noticeable in rear-engine airliners, especially if the toilets are also at the rear.

(b) Apart from seats near the toilet or galley, or in the rear of a rear-engine aircraft, there is little significant noise level differences anywhere inside the cabin.

(c) Noise levels vary significantly throughout the cabin and are highest near toilets and galleys.

(Tick one)

(14) (a) There is no need to make a large aircraft any quieter since 400 passengers, talking at the same time, make more noise than the mechanical sounds.

(b) The mechanical noise inside an aircraft cabin is similar to that made by 400 passengers all talking at once.

(c) In the cabin of a large jet, the sound of 400 passengers talking at the same time is dispersed by the mechanical noises.

(Tick one)

(15) List below

(a) The amenities provided for economy class cabin comfort, and (b) the features provided for passenger safety.

Comfort: adjustable backrests, ...

Safety: crash-resistant seating, ...

II. Study the text, then answer the following according to the instructions.

(1) (a) In spite of all possible precautions, crew training, emergency equipment, passenger good sense, etc, accidents are inevitable.

(b) It is always possible that an accident will happen, however well trained the crew, or efficient the emergency equipment and sensible the passengers.

(c) Accidents are survivable, but only as a result of crew training, aircraft equipment and passenger good sense.

(Tick one)

(2) (a) Being aware of the possibility of an emergency is the best way of surviving it.

(b) Passengers who have been told about emergencies stand the best chance of surviving them.

(c) Preparing one's mind for the possibility of an emergency gives one a better chance of survival.

(Tick one)

(3) (a) High altitude depressurization is so rapid that the chances of getting the oxygen mask on in time are small.

(b) Sudden depressurization at high altitudes leaves one with very little time to familiarize oneself with how to put on the oxygen mask.

(c) Once depressurization occurs at high altitudes it will be too sudden for passengers to do anything about it.

(Tick one)

(4) (a) Since passengers will be required to operate the emergency doors, opening instructions are given in one or more languages.

(b) Instructions for opening emergency exits are given in more than one language so that passengers, if required to do so, can understand them.

(c) Only those passengers who understand the language of the emergency door operation instructions will be able to read them.

(d) As it is possible in an emergency for passengers to operate the exits, instructions on their operation are given in more than one language.

(Tick one)

(5) Severe air turbulence can cause injuries to passengers in flight.

Below is the list (a-d) of the various categories of passengers at risk. In the brackets [ ] opposite each, write the numbers 1 — 4 to, indicate a descending order of risk; i. e place: #4 in the brackets [4] against that category which is least at risk, [1]against that which is most vulnerable to injury.

[ ] (a) passengers walking about at the front of the aircraft.

[ ] (b) seated passengers.

[ ] (c) passengers walking to and from the rear toilets.

[ ] (d) passengers seated, with safety belts fastened.

(6) It may be assumed from the whole text and lists of emergency equipment carried by aircraft that:

(a) most accidents are survivable.

(b) survivable accidents only occur at sea.

(c) accidents invariably cause injury to passengers.

(d) accidents, even survivable ones, often involve fire and fumes.

(e) survivable accidents rarely occur in mid air.

(Tick whichever is/ are correct)

(7) On the left below is a list of various items of emergency equipment; on the right is a list of possible uses. Match the equipment to the uses by writing the appropriate number of list B in the brackets [ ] opposite the items of list

A. [3] has been done for you.

A.	B.
[ ] asbestos gloves	1. for sliding down to safety
[ ] axes	2. for signalling by sunlight
[ ] CO_2 fire extinguishers	3. for scooping water out of a lift raft
[ ] drogue	4. for keeping warm in conditions of extreme cold
[ ] heliograph	5. as a loudspeaker (hailer)
[3] baler (bailer)	6. as eye protection against smoke

[ ] megaphone   7. for preventing a life-raft from drifting from the accident site

[ ] smoke goggles   8. for handling burning materials

[ ] chutes   9. for putting out fires involving electricity

[ ] polar suit   10. for forcing exits

Extensive Reading

Galleys in A380

Galleys are installed in the two deck cabins, to store and prepare food and drinks for the passengers and crew, and to store waste material. Galley can be wet or dry. Wet galleys are used to store and prepare food and drinks; they are connected to potable and waste water systems, to the air extraction and cabin ventilation system and to the power supply. Dry galleys are not connected to any system and are used storage. The number and the location of the galleys depend on the aircraft configuration. The galleys are cooled through a centralized cooling system, called the supplemental cooling system (ATA 21 – 57).

Lavatories in A380

Lavatories are installed in the two decks. They are connected to the vacuum waste system. The number and the location of the lavatories depend on the aircraft configuration.

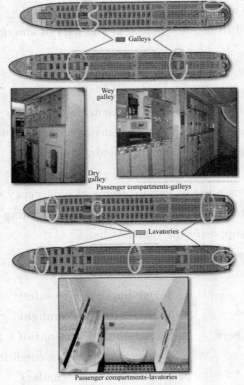

Figure 15 – 9

Lesson 16 Structural Repair Manual

INTENSIVE READING

B737 - 800 STRUCTURAL REPAIR MANUAL INTRODUCTION

General

This Structural Repair Manual gives general data and special instructions for the repair of the Boeing 737 - 800 and 737 - 800 BBJ airplane structure. This manual gives general airplane data, usual procedures, and repair materials. This manual also includes material identification, allowable damage, and repair data for the airplane structure. Procedures usually done together with the structural repair (such as an airplane symmetry check or support of the airplane in the jigged position) are also given. This manual was made as specified in the Air Transport Association Specification 100-Specification for Manufacturer's Technical Data. The data in this manual which has an important effect on the airplane structure is approved by the Federal Aviation Administration.

Some structural components that can be repaired are not included in this manual. It is possible that repairs are not given because a general repair was not easily applied to a specified structure. Also, it is possible that service experience has not shown that a repair is necessary for a specified structure. For some types of damage, an analysis by The Boeing Company is necessary before an applicable repair can be made. When an applicable repair is not found in this manual, look at these items:

1. Do a repair as specified in the general repair practices and procedures given in Chapter 51 of this manual.

2. Prepare a special repair for damage not included in this manual.

3. Replace the damaged part.

This manual will change to include more approved repairs when the service experience shows a necessary repair for a specified structure.

For the most important structural damage, it is recommended that the data in this manual be used together with aid from the Boeing Customer Support Representative. Refer to Paragraph 4 for instructions on how to use the data in this manual to make an order for structural spare parts.

Manual Arrangement and Number System

General

The standard number system for Structural Repair Manuals is given in the Air Transport Association Specification 100 (ATA 100)-Specification for Manufacturer's Technical Data. This number system has three elements with two numbers in each element. The first element identifies the Chapter, the second element identifies the Section, and the third element identifies the Subject. Chapter numbers identify the fuselage, the wing, or other such large airplane structures. Section numbers identify the fuselage section 41, the wing center section, or other such smaller parts. Subject numbers identify the skin, the stringers, or other basic structural elements. Refer to Standard Number System, Figure16 - 1 for a usual Chapter-Section-Subject number as follows.

Chapter Numbers

The Chapter numbers and titles given in Table 16 - 1 identify the data included in this manual.

Table 16 - 1

SRM Chapter numbers and titles	
Chapter number	Tutke
51	Standard Practices and Structures-General
52	Doors
53	Fuselage
54	Nacelles/Pylons
55	Stabilizers
56	Windows
57	Wings

Section Numbers

1. The data in the Chapters is divided into specified Section numbers. Important structural differences between models are identified by special Section numbers.

2. Chapter 51 gives data for those subjects which are applicable to all of the airplane structure. Standard practices, general procedures, typical repairs, and other data applicable to more than one chapter is included in Chapter 51. The Chapter-Section numbers and titles

given in Table 16 – 2 identify the data included in Chapter 51.

Table   16 – 2

Chapter section number	Title
51 – 00	General
51 – 10	Investigation and Cleanup of Damage and Aerodynamic Smoothness Requirements
51 – 20	Processes and Procedures
51 – 30	Materials
51 – 40	Fasteners
51 – 50	Support of Airplane for Repair and Alignment Check
51 – 60	Control Surface Balancing
51 – 70	Repairs

Chap er-section numbers and titles for chapter 51

Note:Other important groups of data in Chapter 51 are identified with a different fourth number in the Chapter-Section number.

3. Chapters 52 through 57 give data for the primary airplane structure groups. Chapter-Section numbers identify the components included in each group. The Chapter-Section numbers and titles given in Table 16 – 3 identify the data included in Chapters 52 through 57.

Table   16 – 3

Chapter section number	Tutke	Chapter section number	Title
Chapter 52	Doors	Chapter 55	Stabilizers
52 – 00	General	55 – 00	General
52 – 10	Passenger/Crew	55 – 10	Horizontal Stabilizer
52 – 20	Emergency Exit	55 – 20	Elevator
52 – 30	Cargo	55 – 30	Vertical Stabilizer
52 – 40	Service	55 – 40	Rudder
52 – 50	Fixed Interior		
52 – 80	Landing Gear		
52 – 82	Main Landing Gear Wing		
53 – 00	General	56 – 00	General
53 – 10	Section 41	56 – 10	Flight Compartment
53 – 30	Section 43	56 – 20	Passenger Compartment
53 – 40	Section 44	56 – 30	Door

Chapter-section numbers and titles for chapters 52 thru 57

Continue

<table>
<tr><td colspan="4" align="center">Chapter-section numbers and titles for chapters 52 thru 57</td></tr>
<tr><td>Chapter section number</td><td>Tutke</td><td>Chapter section number</td><td>Title</td></tr>
<tr><td>Chapter 53</td><td>Fuselage</td><td>Chapter 56</td><td>Windows</td></tr>
<tr><td>53 – 60</td><td>Section 46</td><td>56 – 40</td><td>Inspection and Observation</td></tr>
<tr><td>53 – 70</td><td>Section 47</td><td></td><td></td></tr>
<tr><td>53 – 80</td><td>Section 48</td><td></td><td></td></tr>
<tr><td>Chapter 54</td><td>Nacelles/pylons</td><td>Chapter 57</td><td>Wings</td></tr>
<tr><td>54 – 00</td><td>General</td><td>57 – 00</td><td>General</td></tr>
<tr><td>54 – 10</td><td>Inlet Cowl-CFM56-7</td><td>57 – 10</td><td>Center Wing</td></tr>
<tr><td>54 – 20</td><td>Fan Cowl-CFM56-7</td><td>57 – 20</td><td>Outer Wing</td></tr>
<tr><td>54 – 30</td><td>Fan Duct Cowl and Thrust Reverser-CFM56-7</td><td>57 – 30</td><td>Wing Tip</td></tr>
<tr><td>54 – 40</td><td>Exhaust-CFM56-7</td><td>57 – 40</td><td>Leading Edge and</td></tr>
<tr><td>54 – 50</td><td>Pylon-CFM56-7</td><td></td><td>Leading Edge Devices</td></tr>
<tr><td></td><td></td><td>57 – 41</td><td>Leading Edge</td></tr>
<tr><td></td><td></td><td>57 – 42</td><td>Leading Edge Slats</td></tr>
<tr><td></td><td></td><td>57 – 43</td><td>Leading Edge Flaps</td></tr>
<tr><td></td><td></td><td>57 – 50</td><td>Trailing Edge and</td></tr>
<tr><td></td><td></td><td></td><td>Trailing Edge Devices</td></tr>
<tr><td></td><td></td><td>57 – 51</td><td>Trailing Edge</td></tr>
<tr><td></td><td></td><td>57 – 53</td><td>Trailing Edge Flaps</td></tr>
<tr><td></td><td></td><td>57 – 60</td><td>Ailerons</td></tr>
<tr><td></td><td></td><td>57 – 70</td><td>Spoilers</td></tr>
</table>

Note: Other important groups of data in Chapters 52—57, such as optional engines, are identified with a different fourth number in the Chapter-Section number.

Subject Numbers

1. In Chapter 51 the data included in a Chapter-Section is divided into Subject numbers given in a sequence.

2. In Chapters 52 through 57 the data included in a Chapter-Section is divided into specified Subject numbers. A given Chapter-Section will only use those subjects applicable to

the structure included in that Chapter-Section of the manual. The Subject numbers and titles given in Table 16 – 4 identify the data included in Chapters 52 through 57.

Table   16 – 4

Subject numbers and titles for chapters 52 thru 57			
Chapter section subject number	Title	Chapter section subject number	Title
5X-XX-00	General	5X-XX-50	Floor Panels
—01	Skins and Plates	—51	Floor Structure
—02	Structure Complete	—52	Seat Tracks
—03	Stringers and Skin Stiffeners	—53	Cargo Compartment Special Structure
—04	Intercostals		
—05	Longerons		
—06	Formers	—54	
—07	Frames	to	Not Used
—08	Bulkheads	—69	
—09	Ribs	—70	Fillets and Fairings
—10	Spars		Skins and Plating
—11	Intermediate, Auxiliary and Subspars	—71	Fillets and Fairings, Structure
—12	Keel Structure	—72	Radomes and Antenna
—13	Beams		Fairings
—14	Landing Gear Support	—73	
—15	Door Surround Structure	to	Not Used
—16	Edge Members	—89	
—17 to —29	Not Used	—90 —91 to	Attachment Fittings Not Used
—30	Auxiliary	—99	
—31 to —49	Not Used		

Note: Attachment fittings identified in —90 are fittings with a lug (with a bushing or bearing), links, braces, and hinge fittings (with a bushing or bearing). Other fittings (backup, structural) such as door stop fittings, jack fittings, and splice fittings are identified in the applicable Subject (Example: —10 for spars).

Topics, Figure Numbers, and Page Numbers

1. In Chapter 51 the data included in a Chapter-Section-Subject is divided into topics with figure numbers and page numbers given in Table 16 – 5 below.

Table 16 - 5

Topic, figure, and page numbers for chapter 51

Chapter 51 topic	Figure number	Ppagenumber
General information	Figure 1 - 99	Page 1 - 99
Repair general	Figure 201 - 999	Page 201 - 999
Repair 1, 2, 3, ...	Figure 201 - 999	Page 201 - 999

(2) In Chapters 52 through 57 the data included in a Chapter-Section-Subject is divided into topics with figure numbers and page numbers given in Table 16 - 6 below.

Table 16 - 6

Topic, figure, and page numbers for chapters 52 thru 57

Chapter 52—57 topic	Figure number	Page number
General information	Figure 1 - 99	Page 1 - 99
Identification general	Figure 1 - 99	Page 1 - 99
Identification 1, 2, 3, ...	Figure 1 - 99	Page 1 - 99
Allowable damage General	Figure 101 - 199	Page 101 - 199
Allowable damage 1, 2, 3, ...	Figure 101 - 199	Page 101 - 199
Repair general	Figure 201 - 999	Page 201 - 999
Repair 1, 2, 3, ...	Figure 201 - 999	Page 201 - 999

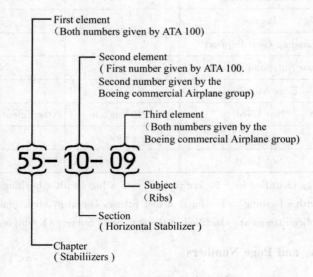

Figure 16 - 1 Standard number system

How to Use the Manual

Cantion: Make sure that you do all of the steps to the end of the procedure. Large blank spaces can occur at the bottom of pages which do not always indicate that you are at the end of the procedure. If you do not make sure that you have completed the procedure, damage to equipment or system malfunction could ocur.

Identify the material of the damaged part. Refer to the applicable material identification data in this manual. For the example used before, in 55 – 10 – 09, the topic "Identification 1 –Horizontal Stabilizer Forward Box Ribs" gives the identification data for the damaged part (Refer to Figure 16 – 2).

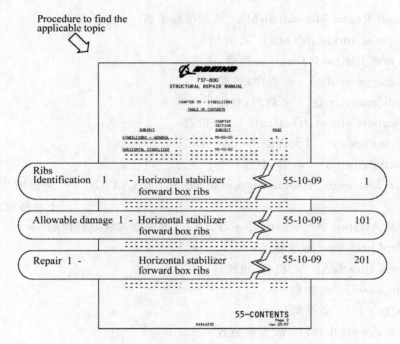

Figure 16 – 2　How to use the manual

Note: The material identification tables and some of the figures give reference drawing numbers. Refer to these reference drawings when the applicable structure is not identified.

Find the allowable damage data for the damaged structure. Refer to the Manual Arrangement and Number System in the Introduction of this manual. Also refer to the List of Chapters for this manual and to the Table of Contents in the applicable chapter. As an example, if the damaged part was a rib in the forward box of the horizontal stabilizer, the allowable damage data is found as follows:

1. The List of Chapters for this manual shows that data on Stabilizers is included in Chapter 55.

2. The Table of Contents for Chapter 55 shows that all data on the Horizontal Stabilizer

is included in the Chapter-Section 55 – 10 – XX. The data on Ribs is given in Chapter-Section-Subject 55 – 10 – 09. The topic "Allowable Damage 1 – Horizontal Stabilizer Forward Box Ribs" gives the allowable damage data for the damaged part.

Find the repair data for the damaged structure. For the example used before, in 55 – 10 – 09, the topic "Repair 1 – Horizontal Stabilizer Forward Box Ribs" gives the repair data for the damaged part.

Unless specified differently, all dimensions in this manual are given in inches.

New Words/Phrases/Expression

1. Structural Repair Manual(SRM)　结构维修手册
2. introduction[ˌɪntrəˈdʌkʃ(ə)n]　n.介绍
3. instruction[ɪnˈstrʌkʃ(ə)n]　n.指令;说明
4. procedure[prəˈsiːdʒə]　n.程序;规程
5. materials[məˈtɪərɪəlz]　n.材料;材质
6. identification[aɪˌdentɪfɪˈkeɪʃ(ə)n]　n.识别
7. allowable damage　可允许损伤
8. symmetry[ˈsɪmɪtrɪ]　n.对称(性)
9. The Air Transport Association Specification 100（ATA100）　美国航空运输协会第100号规范
10. Federal Aviation Administration（FAA）　美国联邦航空管理局
11. standard practice　标准施工
12. fuselage[ˈfjuːzəlɑːʒ]　n.机身(飞机)
13. nacelle[nəˈsel]　n.短舱
14. pylon[ˈpaɪlən]　n.吊架
15. stabilizer[ˈsteɪbɪlaɪzə]　n.安定装置
16. aerodynamic[ˌɛərəʊdaɪˈnæmik]　adj.空气动力学的
17. aerodynamic smoothness requirement　气动光滑性要求
18. fastener[ˈfɑːsnə(r)]　n.紧固件
19. alignment[əˈlaɪnm(ə)nt]　n.校准
20. control surface balancing　操纵面配平
21. emergency[ɪˈmɜːdʒ(ə)nsɪ]　adj.紧急的;备用的
22. cargo[ˈkɑːgəʊ]　n.货物
23. horizontal[hɒrɪˈzɒnt(ə)l]　adj.水平的
24. elevator[ˈelɪveɪtə]　n.升降舵
25. vertical[ˈvɜːtɪk(ə)l]　adj.垂直的
26. rudder[ˈrʌdə]　n.飞机方向舵
27. landing gear　起落架;起落装置

28. inlet cowl　进气整流罩、进气罩

29. thrust reverser　反推装置

30. exhaust[ɪgˈzɔːst；eg-]　n. 排气；排气装置

31. leading edge slat　前缘缝翼

32. trailing edge flap　后缘襟翼

33. aileron[ˈeɪlərɒn]　n. 副翼

34. spoiler[ˈspɒɪlə]　n. 扰流板

35. longeron[ˈlɒn(d)ʒərɒn]　n.（飞机的）纵梁

36. former[ˈfɔːmə]　n. 模型，样板

37. frame[freɪm]　n. 框架

38. bulkhead[ˈbʌlkhed]　n. 隔板

39. rib[rɪb]　n. 翼肋

40. fillet[ˈfɪlɪt]　n. 嵌条；细带；搭接带

41. fairing[ˈfeərɪŋ]　n. 整流罩

42. spar[spɑː]　翼梁

43. keel[kiːl]　n. 龙骨；龙骨脊

44. radome[ˈreɪdəʊm]　n.［雷达］雷达天线罩；整流罩

45. antenna[ænˈtenə]　n.［电讯］天线

46. beam[biːm]　n. 横梁

47. malfunction[mælˈfʌŋ(k)ʃ(ə)n]　n. 故障；失灵

Notes

(1) The data in this manual which has an important effect on the airplane structure is approved by the Federal Aviation Administration.

分析：由 which 引导的定语从句；be approved by 由……批准。

翻译：这本手册里资料由美国联邦航空管理局批准，对飞机结构有重要影响。

(2) Large blank spaces can occur at the bottom of pages which do not always indicate that you are at the end of the procedure.

分析：which 在从句中作主语。

翻译：在页面的底部可能会出现大的空白空间，这并不总是表明你到了程序的结束。

Exercises

Ⅰ. **Choose the best answer according to the figure 16 – 3.**

(1) A. spoiler　　　　　B. rudder　　　　　C. pylon　　　　　D. radome

(2) A. door　　　　　　B. fuselage　　　　　C. window　　　　D. wing

(3) A. door　　　　　　B. fuselage　　　　　C. window　　　　D. wing

(4) A. leading edge slats   B. leading edge flaps

     C. trailing edge flaps   D. spoiler

(5) A. leading edge slats   B. leading edge flaps

     C. trailing edge flaps   D. spoiler

(6) A. elevator   B. wing tip   C. aileron   D. spoiler

(7) A. elevator   B. wing tip   C. aileron   D. spoiler

(8) A. leading edge slats   B. leading edge flaps

     C. trailing edge flaps   D. aileron

(9) A. horizontal stabilizer   B. vertical stabilizer

     C. leading edge slats   D. trailing edge flaps

(10) A. spoiler   B. rudder   C. aileron   D. elevator

(11) A. spoiler   B. rudder   C. aileron   D. elevator

(12) A. horizontal stabilizer   B. vertical stabilizer

       C. leading edge slats   D. trailing edge flaps

(13) A. spoiler   B. rudder   C. aileron   D. elevator

(14) A. spoiler   B. rudder   C. pylon   D. radome

(15) A. pylon   B. fuselage   C. radome   D. wing

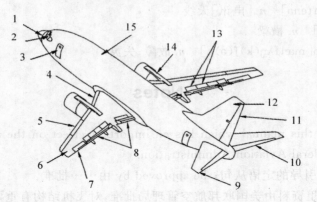

Figure   16 - 3

Ⅱ. Fill in the following blanks according to the text.

(1) The standard number system for Structural Repair Manuals is given in the _____ Specification for Manufacturer's Technical Data.

(2) The standard number system for SRM has _____ elements with _____ numbers in each element.

(3) The first element of the standard number system identifies the _____, the second element identifies the _____, and the third element identifies the _____.

(4) The third element of the standard numbe are given by _____.

Ⅲ. Refering to the SRM introduction, fill in the following blanks.

Find the applicable topic	Chapter-Section-Subject	PAGE NUMBER
(1) 垂直安定面前翼梁腹板可允许损伤鉴定	【　】-【　】-【　】	Page【　】-【　】
(2) 左大翼翼肋的材料、厚度和图号	【　】-【　】-【　】	Page【　】-【　】
(3) 飞机机身 46 段地板梁的材料	【　】-【　】-【　】	
(4) 飞机通用去除腐蚀损伤的工艺规程	【　】-【　】- ××	Page【　】-【　】
(5) 飞机铝合金件采用阿洛丁处理方法	【　】-【　】- ××	
(6) 飞机紧固件安装与拆除	【　】-【　】- ××	
(7) 飞机前登机门外蒙皮材料、厚度	【　】-【　】- ××	
(8) 原材料为 CLAD 2024 - T42,其可替换的材料牌号	【　】-【　】- ××	
(9) 在机身两根桁条之间蒙皮损伤的填平修理	【　】-【　】-【　】	Page【　】-【　】
(10) 机身 43 段龙骨梁下腹板损伤的修理工艺	【　】-【　】-【　】	
(11) 薄壁钣金件上小凹坑的修理工艺	【　】-【　】- ××	Page【　】-【　】
(12) 机身 43 段(站位 STA400 处)加强隔框组件的材料	【　】-【　】-【　】	

EXTENSIV EREADING

MAINTENANCE PUBLICATION

Maintenance FARs

The primary regulatory tools of the Federal Aviation Administration (FAA) are the Federal Aviation Regulations (FARs). To help organize, the FARs are broken down into separate sections, or parts. For example, FAR Part 65 prescribes the requirements, privileges, and limitations for certification of airmen other than flight crewmembers, which includes aviation maintenance technicians.

Copies of the FARs are available from the Government Printing Office or from a number of private suppliers, including the publisher of this textbook. Since the regulations change frequently, all copies must be periodically updated. Repair stations, aviation maintenance training schools, and others affected by the FARs are required to keep their FARs updated. For your course of study, you may use an FAR textbook produced by the government or some other publisher. These publications serve as a good general guide to the content of the regulations. However, when you take your practical test, you must have access to a current set of updated FARs.

Many regulations do not affect the maintenance technician and require no discussion. Others, however, are of vital importance to technicians in the performance of their duties. It is imperative for all technicians to be familiar with these regulations, and to follow them when exercising the privileges of an A&P certificate. Some of the regulations concerning aircraft maintenance and inspection are listed below:

1. FAR Part 01, Definitions and Abbreviations
2. FAR Part 13, Investigation and Enforcement Procedures
3. FAR Part 21, Certification Procedures for Products and Parts
4. FAR Part 23, Airworthiness Standards. Normal Utility and Acrobatic aircraft
5. FAR Part 25, Airworthiness Standards, Transport Category Airplanes
6. FAR Part 27, Airworthiness Standards, Normal Category Rotorcraft
7. FAR Part 33, Airworthiness Standards: Aircraft Engines
8. FAR Part 35, Airworthiness Standards: Propellers
9. FAR Part 39, Airworthiness Directives
10. FAR Part 43, Maintenance, Preventive Maintenance, Rebuilding and Alterations
11. FAR Part 45, Identification and Registration Markings
12. FAR Part 47, Aircraft Registration
13. FAR Part 65, Certification: Airmen other than Flight Crewmembers
14. FAR Part 91, General Operating and Flight Rules
15. FAR Part 121, Certification and Operations: Domestic, Flag, and Supplemental Air Carriers and Commercial Operators of Large Aircraft
16. FAR Part 125, Certification and Operations: Airplanes having a seating capacity of 20 or more passengers, or a maximum pay load capacity of 6,000 lbs or more
17. FAR Part 127, Certification and Operation of Scheduled Air Carriers with Helicopters
18. FAR Part 135, Air Taxi Operators and Commercial Operators
19. FAR Part 137, Agricultural Aircraft Operators
20. FAR Part 145, Repair Stations
21. FAR Part 147, Aviation Maintenance Technician Schools
22. FAR Part 183, Representatives of the Administrator

FAR PART 39 — Airworthiness Directives

When an unsafe condition exists with an aircraft, engine, propeller, or accessory, the FAA issues an Airworthiness Directive (AD) to notify concerned parties of the condition and to describe the appropriate corrective action. No person may operate an aircraft to which an AD applies, except in accordance with the requirements of that AD. AD compliance is mandatory, and the time in which the compliance must take place is listed within the AD.

Information provided in an Airworthiness Directive is considered approved data for the purpose of the AD. The compliance record for ADs must be entered into the aircraft's permanent records.

Airworthiness Directives are issued biweekly. The biweekly listings are published for small general aviation aircraft and accessories in one volume, while the larger aircraft and their accessories are published in a separate volume. This separation of different aircraft categories provides operators with a much simpler means of filing ADs.

ADs are listed by a six digit numerical number. The first two digits denote the year an AD is issued. For example, all ADs issued during, the year 1996 begin with the number 96-. The third and fourth digits of the AD number denote the biweekly issue in which the AD was first published. There are twenty-six issues of the biweekly AD listing issued each year, and the issues are numbered beginning with number01. The last two digits indicate the number of the AD in the specified biweekly listing. For example, the fourth AD issued in the first biweekly publication in May 1996 would be issued the number: 96 - 10 - 04.

Advisory Circulars (ACs)

Many of the technical publications and regulations issued by the FAA are complex in nature and often require additional explanation. As a result, the FAA issues Advisory Circulars (ACs) to inform, explain, and provide further guidance. Advisory circulars are informative only and cannot be used as approved data unless incorporated into a regulation or an airworthiness directive. Advisory circulars are listed in a numerical sequence closely following the same subject areas covered by the FARs.

One of the most popular Advisory Circulars for maintenance technicians is in the AC43 series. AC43. 13 - 1B and — 2A, Acceptable Methods, Techniques and Practices is a highly technical publication covering most of the aircraft maintenance areas which the A&P technician must service. It contains information on standard

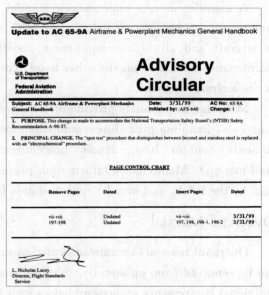

Figure 16 – 4   AC

hardware and torque values, acceptable repair methods, and inspection techniques. (Figure 16 – 4)

Manufacturer's Publications

Aircraft manufacturers provide various manuals with their products to assist technicians in inspection, maintenance, and repair. With few exceptions, manufacturer's manuals are acceptable data. The technician must use manufacturer's maintenance manuals when performing maintenance.

ATA Specification 100

The Air Transport Association of America (ATA) issued specifications for the organization of Manufacturers Technical Data. The ATA specification calls for the organization of an aircraft's technical data into individual systems which are numbered. Each system also has provisions for subsystem numbering. For example, all of the technical information on the Fire Protection system has been designated as Chapter 26 under the ATA 100 specifications, with fire detection equipment further identified by the sub-chapter number 2610, and fire extinguishing equipment as 2620. Because of this specification, maintenance information for all trans-port aircraft is arranged in the same way.

General aviation aircraft manufacturers are in the process of standardizing their maintenance information and ATA Specification 100 will be used as the format for this standardization.

Maintenance Manuals (MM)

A manufacturer's maintenance manual is the primary reference tool for the aviation maintenance technician working on aircraft. Airframe maintenance manuals generally cover an aircraft and all of the equipment installed on it when it is in service. Powerplant maintenance manuals, on the other hand, cover areas of the engines that are not dealt with in the airframe manual.

Maintenance manuals provide information on routine servicing, system descriptions and functions, handling procedures, and component removal and installation. In addition, these manuals contain basic repair procedures and troubleshooting guides for common malfunctions. Maintenance information presented in these manuals is considered acceptable data by the FAA, and may be approved data for the purpose of major repairs and alterations.

Overhaul Manual

Overhaul manuals contain information on the repair and rebuilding of components that can be removed from an aircraft. These manuals contain multiple illustrations showing how individual components are assembled as well as list individual part numbers.

Illustrated Parts Catalog (IPC)

Parts catalogs show the location and part numbers of items installed on an aircraft. They contain detailed exploded views of all areas of an aircraft to assist the technician in locating parts. Illustrated parts catalogs are generally not considered acceptable data for maintenance and repair by the FAA.

Wiring Manuals

The majority of aircraft electrical systems and their components are illustrated in individual wiring manuals. Wiring manuals contain schematic diagrams to aid in electrical system troubleshooting. They also list part numbers and locations of electrical system components.

Structural Repair Manuals (SRM)

For repair of serious damage, structural repair manuals are used. These manuals contain detailed information for repair of an aircraft's primary and secondary structure. The repairs described in a structural repair manual are developed by the manufacturer's engineering staff, and thus are usually considered approved data by the FAA.

Service Bulletins and Notes (SB)

One way that manufacturers communicate with aircraft owners and operators is through service bulletins and service notes. Service bulletins are issued to inform aircraft owners and technicians of possible design defects, modifications, servicing changes, or other information that may be useful in maintaining an aircraft or component. On occasion, service bulletins are made mandatory and are incorporated into airworthiness directives to correct an unsafe condition (Figure 16 – 5).

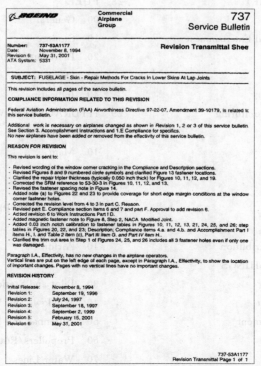

Figure 16 – 5   SB

ATA Specification No. 100

00   General

10   Unassigned

20   Unassigned

30   Recorders

40   Central Computers

50   Central Warning System

32   LANDING GEAR

00   General

10   Main Gear

20   Nose Gear/Tail Gear

30   Extension & Retraction, Level Switch

40   Wheels and Brakes

50   Steering

60   Position, Warning & Ground Safety Switch

70   Supplementary Gear/Skis/Floats

33   LIGHTS

00   General

10   Flight Compartment & Annunciator Panels

20   Passenger Compartments

30   Cargo and Service Compartments

40   Exterior Lighting

50   Emergency Lighting

34   NAVIGATION

00   General

10   Flight Environment Data

20   Attitude and Direction

30   Landing and Taxiing Aids

40   Independent Position Determining

50   Dependent Position Determining

60   Position Computing

35   OXYGEN

00   General

10   Crew

20   Passenger

30   Portable

36   PNEUMATIC

00   General

10   Distribution

20   Indicating

37   VACUUM/PRESSURE

00   General

10   Distribution

20   Indicating

38   WATER/WASTE

00   General

10   Potable

20   Wash

30   Waste Disposal

40   Air Supply

39   ELECTRICAL/ELECTRONIC PANELS AND MULTIPURPOSE COMPONENTS

00   General

10   Instrument & Control Panels

20   Electrical & Electronic Equipment Racks

30   Electrical & Electronic Junction Boxes

40   Multipurpose Electronic Components

50   Integrated Circuits

60   Printed Circuit Card Assemblies

49   AIRBORNE AUXILIARY POWER

00   General

10   Power Plant

20   Engine

30   Engine Fuel and Control

40   Ignition/Starting

50   Air

60   Engine Controls

70   Indicating

80   Exhaust

90   oil

51   STRUCTURES

00   General

52   DOORS

00   General

10   Passenger/Crew

20   Emergency Exit

30   Cargo

40   Service

50   Fixed Interior

60   Entrance Stairs

70   Door Warning

80   Landing Gear

53   FUSELAGE

00   General

10   Main Frame

20   Auxiliary Structure

30   Plates/Skin

10　Power Control	80　STARTING
20　Emergency Shutdown	00　General
77　ENGINE INDICATING	10　Cranking
00　General	81　TURBINES (RECIPROCATING ENG)
10　Power	00　General
20　Temperature	10　Power Recovery
30　Analyzers	20　Turbo-Supercharger
78　ENGINE EXHAUST	82　WATER INJECTION
00　General	00　General
10　Collector/Nozzle	10　Storage
20　Noise Suppressor	20　Distribution
30　Thrust Reverser	30　Dumping & Purging
40　Supplementary Air	40　Indicating
79　ENGINE OIL	83　REMOTE GEAR BOXES (ENG DR)
00　General	00　General
10　Storage (Dry Sump)	10　Drive Shaft Section
20　Distribution	20　Gearbox Section
30　Indicating	

New Words/Phrases/Expression

1. regulatory[ˈregjulətəri]　*adj.*管理的；控制的；调整的
2. Federal Aviation Regulations (FAR)　联邦航空条例
3. privilege[ˈprɪvəlɪdʒ]　*n.*特权；优待
4. certification[ˌsɜːtɪfɪˈkeɪʃən]　*n.*认证；证明，保证；检定
5. airmen[ˈɛəmən]　*n.*航空从业员
6. abbreviation[əbriːvɪˈʃ(ə)n]　*n.*缩写；缩写词
7. enforcement[enˈfɔːsm(ə)nt]　*n.*执行，实施；强制
8. airworthiness standard　适航标准
9. acrobatic[ækrəˈbætɪk]　*adj.*杂技的；特技的
10. rotorcraft[ˈrəʊtəkrɑːft]　*n.*旋翼飞机
11. propeller[prəˈpelə]　*n.*螺旋桨；推进器
12. Airworthiness Directive (AD)　适航指令
13. preventive maintenance　预防性维修；定期检修
14. rebuilding[riˈbɪldɪŋ]　*n.*重建；复原
15. alteration[ɔːltəˈreɪʃ(ə)n；-ɒl-]　*n.*修改，改变；变更
16. registration[redʒɪˈstreɪʃ(ə)n]　*n.*登记；注册
17. helicopter[ˈhelɪkɒptə]　*n.*直升飞机
18. Repair Station　维修站
19. Advisory Circular (AC)　咨询通告

20. Maintenance Manuals（MM） 维修手册

21. Overhaul Manual(OM) 翻修手册

22. Illustrated Pars Catalog（IPC） 图解零件目录手册

23. Wiring Manuals 线路手册

24. Service Bulletin（SB） 服务通告

Exercises

Ⅰ. Fill in the following blanks according to the text.

Some of the regulations concerning aircraft maintenance and inspection are listed below：

（1）FAR Part 01，Definitions and Abbreviations.

（2）FAR Part _____，Airworthiness Standards，Transport Category Airplanes.

（3）FAR Part _____，Airworthiness Directives.

（4）FAR Part 43，Maintenance，Preventive Maintenance，Rebuilding and Alterations.

（5）FAR Part _____，Certification：Airmen other than Flight Crewmembers.

（6）FAR Part _____，Certification and Operations：Domestic，Flag，and Supplemental Air Carriers and Commercial Operators of Large Aircraft.

（7）FAR Part _____，Repair Stations.

（8）FAR Part _____，Aviation Maintenance Technician Schools.

Ⅱ. Answer the following questions according to the text.

（1）What is the FAR?

（2）What is the FAR Part 65?

（3）What is the AD，AC，MM，IPC，SRM，SB?

Ⅲ. Translate the following sentences into Chinese.

（1）To help organize，the FARs are broken down into separate sections，or parts.

（2）For example，FAR Part 65 prescribes the requirements，privileges，and limitations for certification of airmen other than flight crewmembers，which includes aviation maintenance technicians.

（3）When an unsafe condition exists with an aircraft，engine，propeller，or accessory，the FAA issues an Airworthiness Directive（AD）to notify concerned parties of the condition and to describe the appropriate corrective action.

（4）As a result，the FAA issues Advisory Circulars（ACs）to inform，explain，and provide further guidance.

Phrases and Words

A

abbreviation	缩写；缩写词	L16 (E)
abrasive action	磨损作用	L7 (I)
abreast with	并列，并肩地，并排；并排的，并肩的	L15 (I)
abuse	滥用	L2 (I)
accepted practice	习惯做法；常例	L11 (I)
accessibility	易接近；可亲；可以得到	L7 (I)
accommodate	容纳	L2 (I)
accommodation	住处，膳宿；调节；和解；预订铺位	L6 (E)
accumulate	累积；积聚；积攒	L7 (I)
accurate	正确的，精确的	L12 (E)
acoustic	听觉的；声学的；音响的	L8 (I)
acrobatic	杂技的；特技的	L16 (E)
acronym	首字母缩略词	L9 (I)
adhesive	黏合剂；胶黏剂；黏着的；带黏性的	L6 (I)
adversely	不利地；逆地；反对地	L11 (E)
Advisory Circular	咨询通告	L16 (E)
aerodynamic	空气动力学的，[航] 航空动力学的	L6 (I)
aerodynamic smoothness requirement	气动光滑性要求	L16 (I)
age hardened	时效硬化	L4 (I)

aid with	用……来帮助	L6（I）
aileron	副翼	L16（I）
Air Force-Navy（AN）	美国空军海军标准	L10（I）
airfoil	翼面；机翼；螺旋桨	L10（I）
airmen	航空从业员	L16（E）
Airworthiness Directive	适航指令	L16（E）
airworthiness standards	适航标准	L16（E）
aisle	过道，通道	L15（I）
alclad aluminum	包铝	L3（I）
alignment	队列，成直线；校准；结盟	L7（I）
Allen screw	内六角螺钉	L2（I）
Allen wrench	内六角扳手	L2（I）
allowable damage	可允许损伤	L16（I）
alterations	修改，改变；变更	L16（E）
aluminum	［化］铝	L3（I）
AN（Army navy）	陆军海军	L9（I）
anchor nut	托板螺母	L9（I）
annealing	退火，退火（anneal 的现在分词）	L4（I）
anodic	［电］阳极的；上升的	L7（E）
antenna	［电讯］天线	L16（I）
anticipate	预期，期望；占先，抢先；提前使用	L7（I）
antifriction	减低或防止摩擦之物，润滑剂；减少摩擦的	L3（I）
anvil	（铁）砧，砧座	L2（E）
applicable	可适用的；可应用的；合适的	L12（I）
application	应用；申请；应用程序	L10（E）
approximately	大约，近似地；近于	L6（E）
apt	易于……的	L4（E）
arithmetic	算术	L1（I）
artificial aging	人工时效	L4（I）

assemble	集合,聚集;装配;收集;集合,聚集	L6 (I)
assortment	分类;混合物	L13 (I)
asterisk	星号	L9 (I)
autoclave	压热器,高压蒸气灭菌器	L14 (I)
automatic punch	自动中心冲	L2 (I)
auxiliary	助动词;辅助者,辅助物;附属机构;辅助的;副的;附加的	L6 (I)
available	可获得的;可找到的	L10 (E)
aviation snip	航空剪	L2 (I)
avionics compartment	电子舱	L15 (I)
awkward	别扭的,难操纵的	L2 (I)

B

babbit	巴比合金,巴氏合金	L3 (I)
backrest	靠背	L15 (I)
back-up	支持,援助	L11 (E)
baffle	挡板	L10 (I)
ball hammer	圆头锤	L2 (I)
barrel	圆筒	L2 (E)
beam	横梁	L16 (I)
bearing strength	承压强度	L10 (I)
bell crank	曲拐;曲柄;直角杠杆;摇臂	L7 (E)
bending load	弯曲负荷;弯曲荷载	L6 (E)
bevel	斜角,倾斜,斜面	L2 (E)
binary	二进制;二元的	L1 (I)
blemish	瑕疵,缺点,污点,不名誉;玷污,损害,弄脏	L8 (I)
blowout	爆裂;喷出;[电]保险丝烧断	L6 (E)
blunt	钝的	L2 (I)
blunt point	平头	L9 (I)

body hammer	打平锤,锤光锤	L2（I）
bolt	螺栓	L9（I）
borescope	管道镜	L8（I）
borne	忍受;负荷	L6（E）
boron	硼	L5（I）
bottom	底部;末端;底部的	L6（E）
box-end wrench	梅花扳手	L2（I）
breakthrough	突破	L9（I）
brine	卤水;浓盐水;海;海水	L4（I）
brittleness	脆性,脆度	L4（E）
bucking bar	顶铁	L11（I）
buckle	扣住,变弯曲,使弯曲;皮带扣,带扣	L7（I）
bulbed	球状的,圆头的	L10（E）
bulkhead	隔板	L16（I）
burrow	探索,寻找,挖掘,挖出;（兔、狐等的）洞穴,地道	L7（E）
bus bar	汇流条,母线	L3（I）
business class seat	商务舱座椅	L15（I）
butt	粗大的一端,靶垛	L12（I）

C

cabin attendant seat	客舱乘务员座椅	L15（I）
cabin layout	客舱布局	L15（I）
calibrate	校准	L2（I）
calipers	弯脚器,测径器,卡规	L2（E）
cant	斜面,伪善之言,黑话,角落;把……棱角去掉;使……倾斜;甩掉	L7（I）
cantilever	悬臂	L6（E）
cape chisel	前扁尖凿,斜刃凿	L2（I）
carbide	碳化物;硬质合金	L4（I）

carbon fiber	碳化纤维,[材]碳素纤维	L6 (I)
carbon steel	碳钢	L3 (I)
carburizing	渗碳,增碳剂,渗碳剂	L4 (I)
cardiac	心脏(病)的,(胃的)贲门的;心脏病患者,强心剂	L15 (I)
cargo	货物	L16 (I)
casehardening	表层硬化,使表面硬化(caseharden 的现在分词)	L4 (I)
castellation	蝶形	L9 (I)
casting alloy	铸造合金	L3 (I)
catastrophic	灾难的;悲惨的;灾难性的,毁灭性的	L6 (I)
cathodic	阴极的,负极的	L3 (I)
caustic	[化]腐蚀性的;[助剂]腐蚀剂,苛性钠	L7 (E)
cavity	腔;洞,凹处	L6 (E)
center punch	中心冲	L2 (I)
centrifugal	离心的	L1 (I)
centripetal	向心的	L1 (I)
ceramic	陶器的,与陶器有关的, 陶瓷的,陶制的;陶瓷制品,陶瓷器,制陶艺术,陶瓷装潢艺术	L5 (I)
certification	认证;证明,保证;检定	L16 (E)
chamfer	去角;挖槽;斜切	L12 (E)
char	把……烧成炭,把……烧焦,烧焦;碳,(尤指家庭的)杂务	L14 (I)
characteristic	特征;特性;特色	L10 (E)
charter flight	包机	L15 (I)
check nut	防松螺母	L9 (I)
chisel	凿子	L2 (I)
chromium	[化]铬	L3 (I)
chuck	卡盘	L2 (I)

circumference	圆周,周长	L1 (I)
circumstance	情况,环境,条件	L2 (I)
clamp	夹紧,夹钳	L2 (I)
classic configuration	经典布局	L15 (I)
classification	分类;类别,等级	L7 (I)
claw hammer	羊角锤	L2 (I)
clean-out	清除,扫荡	L12 (E)
clearance	清除;空隙;间隙	L10 (E)
clevis pin	销钉	L9 (I)
close tolerance bolt	高精度螺栓	L9 (I)
cockpit	驾驶员座舱;战场	L6 (I)
coincide	一致,符合	L2 (E)
cold chisel	冷錾,冷凿	L2 (I)
cold working	冷加工,冷作	L4 (I)
collar	衬套;颈圈	L10 (E)
collision	碰撞;冲突	L7 (I)
combination set	组合角尺(量具)	L2 (E)
compartment	[建]隔间;区划;卧车上的小客房;分隔;划分	L6 (I)
compasses	圆规	L2 (E)
component	组成的,构成的;成分;组件;[电子]元件	L6 (I)
composite	混合成的,综合成的;合成物,混合物,复合材料	L5 (I)
composite materials	复合材料;聚合物复合材料	L6 (I)
concentration	集中,集合	L12 (I)
concentric	同轴的;同中心的	L11 (E)
conduct	管理;引导;表现	L12 (I)
conductivity	传导性,传导率,电导率	L4 (E)
consist of	由……组成;由……构成;包括	L6 (I)

continuity	连续性;一连串;分镜头剧本	L6（E)
contour	轮廓;等高线;周线;电路;概要;画轮廓;画等高线	L6（I)
control surface balancing	操纵面配平	L16（I)
conventional	传统的;常见的;惯例的	L10（E)
copper	铜	L3（I)
corrode	使腐蚀,侵蚀	L5（I)
corrosion	腐蚀;侵蚀;锈蚀;受腐蚀的部位;衰败	L8（I)
corrosion-resistant	耐蚀	L3（I)
corrosive agent	腐蚀剂,腐蚀介质	L3（I)
corrugate	使起皱;成波状	L12（I)
cotter pin	开口销	L9（I)
countersink	埋头孔	L13（I)
countersink cutter	埋头钻	L11（I)
countersunk	埋头孔;埋头的	L10（I)
countersunk screws	埋头螺丝;沉头螺钉	L6（I)
cowling	飞机引擎罩	L4（E)
crack	裂缝,裂纹	L4（E)
crash axe	撞击斧	L15（I)
crazing	银纹;破裂;龟裂;细裂纹	L7（I)
crease	折痕,折缝;起皱弄皱,使起折痕	L7（I)
critical point	相变点,临界点	L4（I)
cross hammer	横头锤	L2（I)
cross point screw driver	十字螺刀	L2（I)
cross-sectional area	横断面积;断面面积	L7（I)
crystalline	透明的;水晶般的;水晶制的	L7（I)
cube	立方体,立方	L1（I)
cure	治愈,药物,疗法,措施;治愈,矫正,解决,消除,被加工处理,固化	L14（I)

curing agent	固化剂	L14（I）
curvature	弯曲,曲率	L12（E）
cushion	座垫,垫子,起保护（缓冲）作用的事物；（用垫子）使柔和,（跌倒或碰撞时）起缓冲作用	L15（I）
cutout	挖去部分；删除部分；开孔	L12（E）
cutting tool	切割工具	L2（I）
cyaniding	氰化	L4（I）
cyclic action	反复作用	L9（I）
cylinder	圆柱体,气缸	L1（I）
D		
damage	损害,毁坏；损伤,损害,损毁,赔偿金	L14（I）
damage tolerance	损伤容限	L7（I）
debond	脱胶	L8（I）
debris	碎片,残骸	L7（E）
deburr	去毛刺；倒角；清理毛刺	L13（I）
deburring	倒角,去除毛刺；清除飞边	L11（I）
decimal	小数；十进制的,小数的	L1（I）
defect	缺点,缺陷；不足之处	L7（I）
defibrillator	（电击）除颤器	L15（I）
definite	明确的,一定的	L12（E）
deflection	偏向；挠曲；偏差	L7（I）
delamination	分层；分叶；层离；起鳞	L8（I）
deleterious	有毒的,有害的	L7（E）
delicate	精巧的,灵敏的,精密的	L2（E）
demagnetize	消磁,使退磁	L8（I）
density	密度；浓度	L4（E）
dermatitis	皮炎	L5（I）
design criteria	设计准则；设计标准	L6（I）
desire	渴望	L2（I）

deterioration	恶化；退化；堕落	L7 (I)
detrimental	有害的；不利的	L4 (E)
developer	开发者；[摄] 显影剂，显像剂	L8 (I)
diagonal	对角线的，斜的	L2 (I)
diameter	直径	L9 (I)
diamond point chisel	棱形凿，金刚石尖头凿	L2 (I)
disassemble	分解	L9 (I)
discoloration	变色；污点	L7 (I)
dissimilar metal corrosion	异类金属腐蚀	L10 (I)
distort	扭曲；使失真；曲解	L11 (E)
distortion	变形；[物] 失真；扭曲；曲解	L7 (I)
dividers	两脚规，分规	L2 (E)
division	分开，区分，除法	L1 (I)
duckbill pliers	鸭嘴钳	L2 (I)
ductility	展延性，柔软性；韧性；塑性	L4 (E)
duplicate	复写，复制，使加倍	L12 (E)
durability	耐久性；坚固；耐用年限	L7 (I)
dye	染料，染色，颜色；染色，给……染色；染上或粘上（颜色）	L8 (I)

E

eddy current	涡流，涡电流	L8 (I)
elastic limit	弹性极限	L3 (I)
elasticity	弹力，弹性	L4 (E)
electrical conductivity	电导率	L3 (I)
electrochemical action	电化学作用	L7 (I)
electrolyte	电解液，电解质；电解	L7 (E)
electrolytically	以电解	L3 (I)
elevator	升降舵	L16 (I)
elliptical	椭圆的；省略的	L6 (I)

elongate	拉长，使延长，使伸长；拉长，延长，伸长；伸长的，延长的	L7（I）
emergency	紧急情况；突发事件；非常时刻；紧急的；备用的	L6（I）
emergency exit	紧急出口	L15（I）
emergency locator transmitter	事故位置发射机	L15（I）
empennage	［航］尾翼，尾部	L6（I）
enclose	围绕；装入；放入封套	L6（E）
encompass	包含	L2（I）
enforcement	执行，实施；强制	L16（E）
engage	啮合、配合	L9（I）
epoxy	环氧树脂；环氧的	L5（I）
equilateral triangle	等边三角形	L1（I）
erosion	磨蚀，侵蚀，腐蚀	L7（I）
escape slide	应急离机滑梯，紧急滑梯	L15（I）
evaluate	评价；估价；求……的值；评价；估价	L7（I）
exfoliation	剥落；剥落物；表皮脱落	L7（I）
exhaust	排气；废气；排气装置	L10（I）
extraneous material	外来杂质	L7（I）
extreme	极端的；极度的；极端；末端	L6（E）
extreme load	极限载荷	L11（I）
extrusion	挤出；推出；赶出；喷出	L6（I）

F

fabricate	制造；装配	L11（E）
facilitate	促进，助长，使容易，帮助	L14（I）
failsafe	破损安全；失效保护	L6（I）
failure	失败；故障；失效；失败者；破产	L6（I）
fairing	整流装置；整流罩	L4（E）
false brinelling	摩擦腐蚀压痕	L7（E）

far back	遥远	L11 (I)
fastener	紧固件	L3 (I)
fatigue	疲劳,疲乏	L3 (I)
fatigue crack	［力］疲劳裂纹	L7 (I)
Federal Aviation Administration (FAA)	美国联邦航空管理局	L16 (I)
Federal Aviation Regulations (FAR)	联邦航空条例	L16 (E)
fiber	纤维;光纤;(织物的)质地;纤维物质	L5 (I)
field rivet	外场铆钉	L10 (I)
filiform corrosion	［涂料］丝状腐蚀	L7 (E)
fillet	嵌条;细带;搭接带	L16 (I)
fillister head screw	有槽圆头螺钉	L9 (I)
fire extinguisher	灭火器	L15 (I)
fireproof	防火的;耐火的；使耐火;使防水	L6 (E)
firewall	防火墙；用作防火墙	L6 (I)
first aid kit	急救箱	L15 (I)
first class seat	头等舱座椅	L15 (I)
first-aid	急救的,急救用的	L15 (I)
flame-resistant material	阻燃材料	L15 (I)
flange	［机］法兰；给……装凸缘	L6 (E)
flap	襟翼	L4 (I)
flashlight	手电筒;闪光信号灯	L8 (I)
flat head screw	平头螺钉	L9 (I)
floating lift-preserver	救生浮筏	L15 (I)
fluctuating stresses	脉动应力;交变载荷	L7 (I)
fluid	液体,流体;流体的,流动的,液体的，易变的,不固定的	L14 (I)
former	模型,样板	L16 (I)

fraction	分数	L1 (I)
frame	框架;结构;设计;建造	L6 (E)
free fit	自由配合	L9 (I)
freeze-thaw cycling	冻融循环	L14 (I)
furnace	熔炉,火炉	L4 (I)
fuselage	机身(飞机)	L16 (I)
fusibility	熔性,熔度	L4 (E)
G		
galley	[船]船上的厨房	L6 (E)
galvanic corrosion	接触腐蚀;电偶腐蚀;电化学腐蚀	L7 (I)
gap	间隙;缺口;差距	L11 (E)
gauge	行距	L11 (I)
gear	齿轮	L1 (I)
general formula	通式;一般公式;一般准则	L11 (I)
general guidelines	一般准则	L7 (I)
geographic location	地理定位;地理位置	L7 (E)
geometry	几何学	L1 (I)
gimlet point	尖头	L9 (I)
gouge	用半圆凿子挖,欺骗;沟;圆凿,以圆凿刨	L7 (I)
graduation	毕业,刻度,分等级	L2 (E)
graphite	石墨,黑铅;铅笔粉	L5 (I)
greasy	油腻的;含脂肪多的;谄媚的	L7 (I)
grinder	[机]研磨机	L13 (I)
grip	夹紧	L9 (I)
grip length	光杆长度,夹紧长度	L9 (I)
groove	凹槽,槽	L2 (E)
grounding strip	接地片,接地带	L14 (I)
H		
hacksaw	钢锯,弓锯	L2 (I)

hail	冰雹	L12（I）
half round file	半圆锉	L2（I）
hammer	锤子,榔头	L2（I）
hand file	手锉,平锉	L2（I）
hardening	淬火;(使)变硬(harden 的现在分词)	L4（I）
hatch	［船］［航］舱口	L6（E）
hazardous	冒险的,有危险的;冒险地,有危险地;冒险,危险	L15（I）
heat treatment	热处理	L4（I）
helicopters	直升飞机	L16（E）
hermaphrodite	具有相反性质的	L2（E）
Hex-head bolt	外六角头螺栓	L9（I）
high-density seating	高密度座位布置	L15（I）
hinge	铰链,关键,转折点;枢要,中枢;用铰链连接，依……为转移，给……安装铰链,(门等)装有蝶铰;戏弄,装肋于	L6（I）
holding tool	夹持工具	L2（I）
hole repair hardware	孔修补五金件	L9（I）
hollow	空的;中空的,空腹的;凹的;虚伪的;洞;山谷;窟窿	L6（E）
holographic	全息的	L8（I）
honeycomb structure	［航］蜂窝状结构	L6（I）
horizontal	水平的;地平线的;同一阶层的;水平线,水平面;水平位置	L6（I）
hot bonder	热黏合机,热焊机	L14（I）
hub	中心;毂;木片	L12（I）
hydraulic	液压的;水力的;水力学的	L6（I）
hydraulical	液压的	L13（I）
hydroxide	［无化］氢氧化物;羟化物	L7（E）

I

icebox rivet	冰盒铆钉	L10（I）
identical	同一的，完全相同的；完全相同的事物	L7（I）
identification	识别	L16（I）
Illustrated Pars Catalog（IPC）	图解零件目录手册	L16（E）
impact	碰撞，冲击，撞击，冲击力	L14（I）
implement	实施，执行，使生效，实现，落实，把……填满；工具，器械，家具，手段，履行	L14（I）
impregnate	灌输；浸透；充满的	L6（I）
inclusion	包含；内含物	L7（I）
inconsistency	不一致；易变	L12（E）
indent	缩进；订货单；凹痕；契约	L11（E）
inertial	惯性的；不活泼的	L6（I）
inevitable	必然的，不可避免的	L11（E）
ingress	进入，进入权，进食	L14（I）
injury	伤害，损害；受伤处	L7（I）
inlet cowl	进气整流罩、进气罩	L16（I）
insert	嵌块	L9（I）
insertion	插入；嵌入；插入物	L7（I）
insidious	阴险的；隐伏的；暗中为害的；狡猾的	L7（E）
inspection	视察，检查	L7（I）
instruction	指令；说明	L16（I）
insulate	使隔离；使绝缘，使隔热	L3（I）
integral	完整的；整体的；必需的；积分；部分；完整	L6（E）
interchange	互换	L2（I）
interference	干扰,冲突；干涉	L10（E）
interference fit	干涉配合	L9（I）
interferometry	干涉测量	L8（I）
intergranular corrosion	晶间腐蚀；粒间腐蚀	L4（I）

interim	临时的,暂时的，中间的，间歇的；过渡时期,中间时期；暂定	L7 (I)
interlocking joint pliers	内锁支点钳	L2 (I)
Internal wrenching bolt	内六方螺栓	L9 (I)
interrupt	中断；打断；插嘴；妨碍打扰；中断	L6 (E)
intersect	横断，横切；贯穿	L12 (I)
interval	间隔时间	L2 (I)
introduction	介绍	L16 (I)
intrusion	闯入，侵扰	L12 (I)
intrusive	闯入的,打扰的，侵入的	L14 (I)
isosceles triangle	等腰三角形	L1 (I)

J

jackscrew	螺旋起重机；起重螺旋；顶丝	L6 (E)
jaw	颚,钳口	L2 (I)
jeopardize	危害；使陷危地；使受危困	L6 (E)
jig	一种夹具	L4 (E)

K

keel	龙骨；龙骨脊	L16 (I)
kinetic	运动的	L1 (I)
knife file	刀型锉,刀锉	L2 (I)

L

lamina	薄板,薄层,叶片	L5 (I)
laminate	（将薄片砌合在一起）制成；锻压成薄片；层压材料；叠层,层压,由薄片叠成的	L5 (I)
landing gear	起落架,起落装置	L3 (I)
laser shearography	激光剪切测量技术；激光错位散斑干涉测量技术	L8 (I)
lateral axis	[机] 横轴,[数] 横轴线	L6 (I)
layout	布局；设计；安排；陈列	L12 (E)

maintenance technician	维修技术员	L11（I）
malfunction	发生故障，不起作用；故障，失灵，疾病	L7（I）
malleability	有延展性，有锻塑性，柔韧性	L4（E）
maneuver	机动，演习	L7（I）
manganese	[化]锰（元素符号 Mn）	L3（I）
manual release tool	手动释放工具	L15（I）
mar	毁坏，损坏，弄糟，糟蹋，玷污；污点，瑕疵，障碍，损伤，毁损	L8（I）
marine environment	[环境]海洋环境	L7（E）
materials	材料；材质	L16（I）
matrix	基质；[数]矩阵；模型；母体；子宫	L5（I）
meal tray	小餐架	L15（I）
megaphone	扩音器	L15（I）
metallic compound	金属化合物	L7（E）
micrometer	测微计，千分尺	L2（E）
Military Standards（MS）	军用标准	L10（I）
mill file	扁锉	L2（I）
milled	磨碎的；滚花的；铣成的	L10（E）
moderately	适度地；中庸地；有节制地	L7（I）
moisture	水分；湿度；潮湿；降雨量	L7（I）
molybdenum	[化]钼	L3（I）
momentary contact	瞬间接触，瞬时接触	L7（I）
Monel	蒙乃尔铜–镍合金	L3（I）
movable joint	活接头	L9（I）
multiplication	乘法	L1（I）
multiply	繁殖，乘，增加	L2（E）

N

nacelle	气球吊篮；飞机的驾驶员室；飞机的引擎机舱	L6（I）

nacelles	短舱（nacelle 的复数形式）	L16（I）
narrow-bodied airliner	窄体客机	L15（I）
national coarse thread	国家标准粗牙螺纹	L9（I）
national fine thread	国家标准细牙螺纹	L9（I）
natural aging	自然时效	L4（I）
needle	针	L2（E）
needle nose pliers	尖嘴钳	L2（I）
negligible	微不足道的，可以忽略的	L7（I）
neutralize	抵销，使……中和，使……无效，使……中立；中和，中立化，变无效	L7（E）
nibbler	切片机；[机]毛坯下料机	L13（I）
nick	刻痕；凹隙	L2（I）
nickel	[化]镍；镀镍于	L3（I）
nitride	氮化物	L4（I）
nitriding	渗氮，表面氮化	L4（I）
nondestructive	非破坏性的	L8（I）
nonmetallic material	非金属材料	L7（I）
nonstructural	不作结构材料的	L6（E）
normalizing	常化，正火；（使）正常化（normalize 的现在分词）	L4（I）
notch	刻痕；凹口	L13（I）
numerous	许多的，无数的	L12（E）
nut	螺母	L9（I）

O

octagonal	八边形的，八角形的	L12（I）
offset	偏置	L2（I）
open-end wrench	开口扳手	L2（I）
optimum performance	最佳操作特性；最佳性能	L11（I）
orientation	方向,定位,取向,排列方向	L14（I）

orthotropic	正交的	L5 (I)
oval	椭圆形的	L9 (I)
over-aged	过时效,过度失效	L4 (I)
overemphasize	过分强调	L7 (E)
Overhaul Manual	翻修手册	L16 (E)
overload	超载,超过负荷;超载,负荷过多	L7 (I)
oxide	[化] 氧化物	L7 (E)
oxygen mask	氧气面罩	L15 (I)

P

parallel	平行线,对比;使……与……平行;平行的, 类似的,相同的	L2 (I)
parallelogram	平行四边形	L1 (I)
part number	件号	L10 (I)
pattern	模式;图案;样品	L12 (I)
pebbles	卵石	L12 (I)
penetrate	穿透,渗透	L4 (I)
penetration	渗透;穿透;突破;洞察力	L8 (I)
perforate	穿孔于,打孔穿透	L12 (I)
permanent distortion	永久变形;残留变形	L7 (I)
permeability	渗透性;磁导率;可渗透性	L8 (I)
permissible limit	容许极限,允许的限度	L7 (I)
pin	销钉、销子	L9 (I)
pin punch	销冲,销冲头	L2 (I)
pitch	间距、间隙	L11 (I)
pivot	枢轴,支点	L2 (E)
plaincarbon steel	普通碳钢,碳素钢	L3 (I)
plain washer	普通垫圈	L9 (I)
plainly	明显地;清楚地;简单地;坦率地	L7 (I)
plastic hammer	塑料锤	L2 (I)

plexiglass	树脂(有机)玻璃	L13 (I)
pliers	钳子	L2 (I)
ply	股,层,厚,(夹板的)层片	L14 (I)
plywood	夹板,胶合板,多层板	L6 (I)
polar kit	极地服,极地用品包	L15 (I)
polymer	多聚物;[高分子]聚合物	L5 (I)
polysulfide	多硫化物	L14 (I)
portable	手提的,轻便的;手提式打字机	L8 (I)
portable oxygen cylinder	便携式氧气瓶	L15 (I)
portion	一部分,一份	L12 (E)
pounding tool	敲击工具	L2 (I)
powdery	粉的;粉状的;布满粉状物的	L7 (E)
power	[数]幂,[机]功率	L1 (I)
precipitation heat treatment	沉析热处理	L4 (I)
predominately	adv. 占优势地;有影响力地;更大量地	L7 (E)
preload	预加载	L9 (I)
prepreg	(塑料或其他合成材料在模塑之前用树脂浸泡的)预浸料坯	L14 (I)
pressure bulkhead	气密框;气密隔板	L6 (E)
pressurize	密封;增压;使……加压,使……压入	L6 (E)
preventive maintenance	预防性维修;定期检修	L7 (E)
preventive maintenance	预防性维修;定期检修	L16 (E)
prick punch	划线冲子,冲孔器中心冲头	L2 (I)
primary	主要的;初级的;基本的;原色;最主要者	L6 (I)
principal	主要的,资本的;首长;校长;资本;当事人	L6 (I)
privilege	特权;优待	L16 (E)
procedure	程序;规程	L16 (I)
propeller	螺旋桨;推进器	L16 (E)
proportion	比例;使成比例	L1 (I)

protectant	保护剂;杀虫剂	L7 (E)
Protective Breathing Equipments (PBEs)	呼吸防护设备	L15 (I)
protective coating	保护涂料;保护层;保护涂层	L7 (I)
protective finish	表面处理,[涂料] 保护涂层	L7 (E)
protruding	突出的;伸出的	L10 (I)
pry	撬动,撬开	L11 (E)
pry bar	撬杆,撬槓	L2 (I)
pulley	滑轮;皮带轮	L1 (I)
pulse-echo	脉冲回波	L8 (I)
punch	冲子	L2 (I)
puncture	刺穿,揭穿,削弱,被刺穿,被戳破;穿刺,刺痕	L7 (I)
pylon	吊架	L16 (I)

Q

quench	将(热物体)放入水中急速冷却	L4 (I)
quenching medium	淬火介质,淬火剂	L4 (I)

R

radiation	辐射;放射物;辐射状;分散	L8 (I)
radiographic	X 光线照相术的	L8 (I)
radome	[雷达] 雷达天线罩;整流罩	L16 (I)
ratchet	棘轮,棘齿	L2 (I)
ream	扩展;挖;铰孔	L10 (E)
reamed hole	铰制孔	L9 (I)
reamer	铰刀	L2 (I)
rear	向后;在后面;后方的;后面的;背面的;后面;屁股;后方部队	L6 (E)
rebuilding	重建;复原	L16 (E)
receptacle	[植] 花托;容器;插座	L6 (I)

rectangle	长方形,矩形	L1（I）
rectangular	矩形的,成直角的	L12（E）
registration	登记;注册	L16（E）
regulatory	管理的;控制的;调整的	L16（E）
reinforce	加强,加固;强化;求援;加强	L6（E）
reinforcement	加强物;增援;补给品;援军	L5（I）
reinforcement	加固;增援;援军;加强	L12（E）
Repair Station	维修站	L16（E）
repairable	可修理的;可挽回的	L12（I）
residue	残渣;剩余;滤渣	L7（I）
resin	树脂,合成树脂,松香;用树脂处理,涂擦树脂于	L5（I）
retaining collar	止动环	L9（I）
retentivity	保持力;[物]顽磁性	L8（I）
rib	肋骨;排骨;肋状物;翼肋	L6（I）
rigidity	[物] 硬度;[力] 刚性;	L6（I）
rivet	铆钉;铆接;固定	L6（I）
roll pin	柱形插销;滚销	L9（I）
rotorcraft	旋翼飞机	L16（E）
round file	圆锉	L2（I）
round head screw	半圆头螺钉	L9（I）
rudder	飞机方向舵	L16（I）
rupture	破裂,决裂,疝气;破裂,发疝气,使破裂,断绝,发生疝	L7（I）

S

safety-belt anchorage	安全带卡头,安全带固定	L15（I）
sandblast	喷沙,喷沙器;喷沙	L14（I）
scarf	嵌接,围巾,领巾,桌巾,台巾,围(围巾),打(领带),披(披巾),用围巾围	L14（I）

scheduled maintenance checklist	定期维护检查单；计划维护清单	L7 (E)
scheduled service/flight	定期航班	L15 (I)
scrape	刮；擦伤	L11 (I)
scratch	乱涂，勾抹掉，擦，刮	L2 (E)
screw	螺钉、螺丝	L9 (I)
screw driver	螺丝刀	L2 (I)
scriber	划线器，描绘标记的用具	L2 (E)
sealant	密封剂	L14 (I)
seam	缝，接缝；缝合，接合，使留下伤痕，裂开，产生裂缝	L7 (E)
seat belt	座位安全带，座椅固定带	L15 (I)
Secondary Structure	次要结构	L6 (I)
second-generation	第二代	L15 (I)
self-locking	自锁	L9 (I)
self-tapping screw	自攻螺钉	L9 (I)
serration	锯齿	L2 (I)
Service Bulletin (SB)	服务通告	L16 (E)
shear lockbolt	抗剪切自锁螺栓	L9 (I)
shear strength	剪切强度	L10 (I)
sheet metal	钣金工	L2 (I)
shock resistance	耐冲击性	L10 (I)
shop	车间	L2 (I)
shroud	覆盖物；[电] 护罩	L10 (I)
shuttle aircraft	短程(穿梭)机	L15 (I)
silicon	[化]硅，硅元素	L3 (I)
silver	银，银币，银制品，银色；银制的，银色的，银白色的；（在某物上）镀银；使具有银色光泽，使变成银色	L3 (I)
sleeve	套管，轴套	L9 (I)

slip joint pliers	鱼口钳	L2 (I)
slotted screw	带槽螺钉(一字螺钉)	L2 (I)
soak	浸，泡，浸透	L4 (I)
socket wrench	套筒扳手	L2 (I)
solder	焊接，锡焊	L2 (I)
solid wrench	呆扳手(开口扳手的别称)	L2 (I)
solid-shank rivet	实心杆铆钉	L4 (I)
sound-proof	隔音的	L6 (E)
spanwise	顺翼展方向的	L6 (I)
spar	翼梁	L6 (I)
special washer	专用垫圈，特殊垫圈	L9 (I)
spectrum	光谱；波谱；范围；系列	L8 (I)
speed handle	快速摇把	L2 (I)
sphere	球体	L1 (I)
spindle	轴，杆，心轴	L2 (E)
spinner	螺旋桨整流罩，桨毂盖	L12 (E)
spiral flute	螺槽	L2 (I)
spoiler	扰流板	L16 (I)
spray	喷雾，喷雾器，水沫；喷射，喷	L7 (I)
sprocket	链轮齿	L3 (I)
square	正方形，平方，直角尺	L1 (I)
squeeze	挤压	L9 (I)
squeeze riveter	压铆机	L11 (I)
stabilizer	稳定器；安定装置	L6 (I)
stagger	交错排列	L12 (I)
staggered	错列的；吃惊的	L11 (I)
stainless steel	不锈钢	L3 (I)
staircase	楼梯，楼梯间	L15 (I)
standard practice	标准施工	L16 (I)

starting punch	起始冲	L2 (I)
stiffen	变硬;变猛烈;变粘;使变硬;使黏稠	L6 (I)
stiffener	加固物;[建]加劲杆;刚性元件	L6 (I)
stiffness	刚度;僵硬;生硬;强直;顽固	L5 (I)
straight hammer	直头锤	L2 (I)
straightedge	(画直线用的)直尺	L11 (E)
strain	应变	L1 (I)
streamline	流线型的	L10 (I)
stress	应力	L1 (I)
stress corrosion	应力腐蚀	L4 (I)
stringer	纵梁,纵桁	L6 (I)
stripper	褪去剂,剥皮剂,脱衣舞表演者,脱衣舞女	L14 (I)
structural integrity	结构完整性	L7 (I)
Structural Repair Manual (SRM)	结构维修手册	L16 (I)
structure screw	结构螺钉	L9 (I)
subject to	遭受	L9 (I)
substantially	本质上,实质上	L9 (I)
substitute	代用品;代替者;替代;代替	L6 (I)
substructure	基础;子结构;底部构造	L7 (I)
subtlety	精细,细微的差别	L2 (I)
sulfate	[无化]硫酸盐;使成硫酸盐,用硫酸处理,硫酸盐化	L7 (E)
surface corrosion	表面腐蚀	L7 (E)
survival kit	救生包,救生装备	L15 (I)
susceptible	易受影响的,易感动的,容许……的;易得病的人	L7 (E)
suspend	延缓,推迟;使暂停;使悬浮;悬浮	L6 (I)
symmetrically	对称地;平衡地;匀称地	L6 (I)

symmetry	对称(性)	L16（I）
T		
tail cone	[航]尾锥；尾锥体；机身末端	L6（I）
tap	轻敲；轻打；装上嘴子	L11（E）
tap testing	敲击检查	L8（I）
taper	锥度	L2（I）
taper pin	锥形销；圆柱销	L9（I）
tempering	回火；使回火（temper 的现在分词）	L4（I）
temporarily	临时地	L10（E）
tensile lockbolt	抗拉伸自锁螺栓	L9（I）
The Air Transport Association Specification 100（ATA100）	美国航空运输协会第 100 号规范	L16（I）
thermal imaging	热成像	L8（I）
thermocouple	热电偶	L14（I）
thermography	热谱；温度记录；自记温度	L8（I）
thermoplastic	热塑性的；热塑性塑料	L5（I）
thermoset	热固树脂，热固塑料；热固的	L5（I）
thimble	顶针，嵌环，套管	L2（E）
thread	螺纹	L9（I）
three-square file	三角锉	L2（I）
threshold	入口；门槛；开始；极限；临界值	L7（I）
thrust reverser	反推装置	L16（I）
tight fit	紧密配合	L9（I）
tin snip	铁皮剪	L2（I）
titanium	[化]钛	L3（I）
toggle	开关，触发器	L2（I）
tolerance	宽容，容忍；限度；公差；耐受性	L9（I）
torque box	扭力盒	L6（E）
torque value	力矩值	L2（I）

torsion bar torque wrench	扭力杆式力矩扳手	L2（I）
toughness	韧性，坚韧，刚性	L4（E）
tourist (coach) class seat	旅客舱(经济舱)座椅	L15（I）
toxic fume	毒烟	L15（I）
trailing edge	(飞机的)机翼后缘	L6（I）
trailing edge flaps	后缘襟翼	L16（I）
transducer	传感器，变频器，变换器	L8（I）
transfer punch	转换冲	L2（I）
transition	过渡	L9（I）
trapezoid	梯形	L1（I）
triangle	三角形	L9（I）
trigonometry	三角学，三角法	L1（I）
trim	修整，配平	L2（I）
truss head screw	大圆头螺钉	L9（I）
tubular	管状的	L6（E）
tungsten	钨	L5（I）
turning tool	拧动工具	L2（I）
turnlock fastener	转锁紧固件	L9（I）
twist drill	麻花钻	L2（I）

U

ultrafine	非常细微的	L7（I）
ultrasonic	超声的，超音波的，超音速的；超声波	L8（I）
underrate	对(某人或某事物)评价过低；看轻；轻视	L8（I）
unidirectional	单向的，单向性的	L5（I）
uniformity	均匀性；一致；同样	L11（E）
uniqueness	独特性；独一无二；单值性	L6（I）
universal	普遍的；通用的	L10（I）
unstable	不稳定的；动荡的；易变的	L7（I）
upholstery	家具装饰业，室内装饰品	L15（I）

V

vacuum	真空，空白，空虚，清洁；用真空吸尘器清扫	L14（I）
vanadium	钒	L3（I）
vector	矢量，向量	L1（I）
velocity	速度	L1（I）
ventilation	通风设备；空气流通	L6（I）
vernier	［机］游尺，游标，游标卡尺	L2（E）
versatile	万向的；万能的	L2（I）
vertical	垂直的，直立的；垂直线，垂直面	L6（I）
vertical fin	垂尾，垂直安定面	L6（E）
V-groove	V 形槽	L2（I）
vibrate	振动	L9（I）
vibration	振动；犹豫；心灵感应	L7（I）
vicinity	邻近，附近；近处	L7（I）
violent	暴力的；猛烈的	L7（I）
viscous	黏性的；黏的	L7（E）
vise grip	大力钳	L2（I）
visual	视觉的，看得见的；光学的画面，图像	L2（I）
visually	形象化地；外表上；看得见地	L7（I）
vixen file	弧纹锉，弯纹锉	L2（I）
volume	体积；音量	L1（I）
vulnerability	易损性；弱点	L6（I）

W

washer	垫片	L9（I）
water-displacing lubricant	排水润滑剂	L7（E）
wear resistance	耐磨性	L4（I）
webbing	带子；边带；厚边	L13（I）
wet lay-up	湿铺层	L14（I）

whisker	须晶，细须	L5 (I)
wide-bodied	宽体的	L15 (I)
wide-body jet	宽体客机	L4 (I)
wing chord	［航］翼弦	L6 (I)
wing tip	（飞机的）翼尖；翼梢	L4 (E)
Wiring Manuals	线路手册	L16 (E)
wood rasp	木锉	L2 (I)
work	功	L1 (I)
work-harden	加工硬化	L11 (I)
woven	编，织，织成（ weave 的过去分词）；编排；杜撰；（把……）编成	L5 (I)
wrap	卷，裹	L2 (I)
wrench	扳手	L2 (I)
wrinkle	皱纹；起皱；使起皱纹	L7 (I)
wrought alloy	可锻合金，形变合金	L3 (I)

Z

zinc	锌	L3 (I)

References

[1] 白杰,张帆. 民航机务英语教程[M]. 北京:中国民航出版社,1997.

[2] 朱敬才,杨爱荣. 民航公共英语教程[M]. 北京:中国民航出版社,1997.

[3] 温丹丽,隋传国. 电专业英语阅读教程[M]. 大连:大连理工大学出版社,2004.

[4] 阎庆甲,阎文培. 科技英语翻译方法(修订版)[M]. 北京:冶金工业出版社,1992.

[5] 揣成智. 高分子材料工程专业英语[M]. 北京:中国轻工业出版社,1999.

[6] 周光炯. 力学与工程科学专业英语[M]. 北京:北京大学出版社,1997.

[7] Fedral Aviation Administration. Airframe & Powerplant Mechanics General Handbook[M]. Washington: United States Government Publishing Office, 1970.

[8] Fedral Avidtion Administration. Airframe & Powerplant Mechanics Airframe Handbook[M]. Washington: United States Government Publishing Office, 1972.

[9] Jeppesen Sanderson Inc. A&P Technician General Textbook[M]. Colorado: Jeppesen Sanderson, 1996.

[10] Jeppesen Sanderson Inc. A&P Technician Airframe Textbook [M]. Colorado: Jeppesen Sanderson, 1996.

[11] Dale Crane. Training Manual General Section Books 1 Through 7, Integrated Training Program For The Aviation Maintenance Technician [M]. Newcastle: Aviation maintenance publisher, Inc. , 1981.

[12] Dale Crane. Training Manual Airframe Section Textbook Chapters One Through Nine, Integrated Training Program For The Aviation Maintenance Technician[M]. Newcastle: Aviation maintenance publisher, Inc. , 1985.

[13] Robert D Smith. Technical Mathematics[M]. Fourth Edition. New Britain: Central Connecticut State University, 2002.

[14] Jeppesen Sanderson Inc. Advanced Composites [M]. Colorado: Jeppesen Sanderson, 1996.

[15] 朱敬才,赵宁,王爱国,等. 民航专业技术职称(资格)英语水平考试指南[M]. 北京:中国民航出版社,1995.

[16] Fedral Aviation Administration. Aircraft Inspection Repair & Alterations, Acceptable Methods, Techniques, and Practices[M]. Washington: Untied States Government Publishing Office, 1998.

[17] Federal Aviation Administration. Aviation Maintenance Technician Handbook – Airframe: Volume 1[M]. Washington: Untied States Government Publishing office, 2012.

[18] C Dorworth, Ginger L Gardiner, Greg M Mellema. Essentials of Advanced Composite Fabrication and Repair[M]. Newcastle: Aviation Supplies & Academics, Inc. 2009.

[19] Dale Crane. Aviation Mechanic Handbook[M]. Sixth Edition. Newcastle: Aviation Supplies & Academics, Inc. 2011.

[20] Alan Baker, Stuart Dutton, Donald Kelly. Composite Materials for Aircraft Structures[M]. Second Edition. Dallas City: American Institute of Aeronautics and Astronautics, Inc. , 2004.